A Field Guide to

REPTILES OF QUEENSLAND

THIRD EDITION

A Field Guide to REPTILES OF QUEENSLAND

THIRD EDITION

STEVE K. WILSON

This edition published in 2022 by Reed New Holland Publishers
Sydney

Second edition published in 2015
First published in 2005

newhollandpublishers.com

A record of this book is held at the National Library of Australia.

ISBN 978 1 92554 684 2

Managing Director: Fiona Schultz
General Manager: Olga Dementiev
Publisher and Project Editor: Simon Papps
Designer: Andrew Davies
Production Director: Arlene Gippert

Front cover: Prickly Knob-tailed Gecko *(Nephrurus asper).* S. Wilson

Back cover, from top to bottom: Blue-throated Rainbow Skink *(Carlia rhomboidalis).* S. Wilson; Green Python *(Morelia viridis).* S. Wilson; Worrell's Turtle *(Emydura worrelli).* S. Wilson; Freshwater Crocodile *(Crocodylus johnstoni).* S. Eipper.

Keep up with Reed New Holland and New Holland Publishers on Facebook
facebook.com/ReedNewHolland
facebook.com/NewHollandPublishers

CONTENTS

ACKNOWLEDGEMENTS

I could not have written this book without the assistance of many people.

For generous access to their photographs, I am grateful to: Kieran Aland, Luke Allen, Michael Anthony, Bob Ashdown, John Augusteyn, Ant Backer, Jessica Bolton, Rohan Clarke, Rob Browne-Cooper, John Cann, Hal Cogger, Patrick Couper, Ross Coupland, Alexander Davies, Jordan de Jong, Lauren Dibben, Col Dollery, Steve Donnellan, Scott Eipper, Angus Emmott, Jules Farquhar, Graeme Gow, Greg Harold, Harry Hines, Paul Horner, Conrad Hoskin, Dave Knowles, Alex Kutt, Col Limpus, Stewart Macdonald, Roy Mackay, Stephen Mahony, Lindley McKay, Mark O'Shea, Arne Rasmussen, Eric Rittmeyer, Lewis Roberts, Dave Robinson, Mark Sanders, Brendan Schembri, Gunther Schmida, Matthew Shaw, Glenn Shea, Andrew Smith, Ruchira Somaweera, Dave Stanton, Gary Stephenson, Rob Steubing, Gerry Swan, Mike Swan, Rob Valentic, Eric Vanderduys, Harold Voris, Anders Zimny and Stephen Zozaya. Unacknowledged photographs are my own.

Martin Fingland of Brisbane Forest Park, Richard Jackson of Australia Zoo, and Terry Adams and Kieran Aland have allowed me to photograph specimens in their care.

Patrick Couper, Andrew Amey and Heather Janetzki of the Queensland Museum's Vertebrate Zoology department provided information on distribution and access to specimens in the museum's collection. Blanche D'Anastasi kindly reviewed the sea snake material.

I have been enriched by the time I have spent in the company of dedicated naturalists and herpetologists who have enthusiastically shared their field skills, their knowledge about reptile taxonomy, behaviour and conservation, and their contagious passion for an often misunderstood group of animals. For this, I would like to thank Loren Appleby, Patrick Couper, Garth and Ross Coupland, Jeannette Covacevich, Greg Czechura, Jo Davis, Angus Emmott, Mark Hanlon, Greg Harold, Tony and Katie Hiller, Rod Hobson, Richard Johnson, Dave Knowles, Brad Maryan, Gerry Swan, Mike Swan, Eric Vanderduys and Anders and Gina Zimny.

My wife Marilyn has been a tremendous support. She ensured that all computer processes ran smoothly, and continues to reinforce my belief that if you cannot master some key skills, marry someone who can. My parents, Joy and Ken, always encouraged me to follow my dreams.

PREFACE TO THE THIRD EDITION

The primary aim of this field guide is to aid identification by comprehensively documenting the growing number of described reptile species occurring in Queensland. It also has other vital roles to play, fostering awareness in our wildlife and natural environments, and stimulating a desire to protect Queensland's priceless wildlife heritage.

Of the 537 reptile species listed in this book, a total of 207 species are endemic to Queensland. They occur nowhere else so their management and ultimate fate lies entirely in our hands, and on activities undertaken within the state borders.

Revised editions like this are needed to keep abreast with taxonomic changes that herald descriptions of new species. It is an ongoing process, and the adding of species to the national and state tally continues to reinforce just how extraordinary is the diversity in this part of the world. This volume is current to 31 December 2021.

Some of the most significant additions are among members of the burrowing skink genus *Lerista* in north-eastern Queensland. Thanks to the application of genetics and morphological studies, the region is now considered a hot-spot in the evolution of these secretive lizards.

Austral geckos of the family Diplodactylidae have also seen a significant rise in numbers, particularly in the eastern interior. They include some spectacular species that are believed to have extremely restricted distributions.

There have been many changes to the Queensland landscape since the first edition of this book was published in 2005. The most obvious is habitat loss. Some of the thriving reptile communities I used to visit, including several localities where images in this book were taken, simply do not exist anymore. They have fallen victim to land clearing, or have been severely degraded by feral animals and exotic plants. Unfortunately these processes are continuing and sometimes accelerating.

Other changes are less obvious, and as a consequence perhaps more insidious. It has become increasingly apparent to me over the past decade, that reptile numbers may be declining in some localities, even in areas where landscapes such as reserves, national parks and state forests appear intact.

I believe the severe heatwaves and protracted droughts of recent years have taken a toll in depleting some reptile populations, and presumably those of other vertebrates and invertebrates. It appears that their numbers have not yet recovered.

It was only a couple of years ago that subtropical rainforests, unburned for millennia, were ablaze in the Scenic Rim. Then thousands of square kilometres of Mitchell Grass Downs lay under a sheet of floodwater. We may be witnessing the negative impacts of climate change on the Queensland environment.

It is my fervent hope that all of the species portrayed in this field guide, and others still undescribed, will continue to play integral roles in diverse Queensland ecosystems beyond the forseeable future. Of course, that depends on how we as the custodians manage the only environment we have.

Steve K. Wilson, December 2021

INTRODUCTION

In a continent famous for the diversity and unique character of its biological resources, Queensland can truly be said to have more than its fair share. With a tropical climate, Australia's most complex rainforests, deserts as harsh as any the continent has to offer, and the world's largest coral reef, the state is a repository of extraordinary natural wealth.

Reptiles are particularly well represented. There are 537 species covered in this book, of which nearly half are endemic to Queensland. Distributions are sometimes extremely restricted, with some species confined to single localities. For example, rainforests in the Central Queensland Coast harbour at least four species of leaf-tailed geckos (*Phyllurus*). None overlap but they may be separated by as little as 20km. Their evolution in fragmented stands of habitat demonstrates the conservation value of seemingly insignificant forest pockets and the antiquity of the narrow barriers that isolate them.

Lizards like these, along with the sea turtles that nest on our beaches, the crocodiles in our northern waterways, the infamous taipan of our tropical woodlands and canefields, and even the skinks skittering along our garden edges, create the unique blend that is Queensland's reptile fauna. Their fates rest on management practices that take place within a state currently leading Australia in its rate of land clearing.

Conservation

Thus far, Queensland reptiles have fared better than mammals and birds since European settlement. They have not yet suffered the same appalling extinction rates, though they do face continuing declines from a number of threats.

One species of burrowing skink, the Retro Slider (*Lerista allenae*), was feared extinct, but in 2010 relict populations were rediscovered clinging to roadside verges near Clermont. Habitat loss has brought them close to the edge and it will take careful management to ensure they survive.

The loss of native vegetation has been the most substantial and urgent problem facing Queensland's reptiles and other fauna. The estimated annual toll from broad-scale clearing between 1997 and 1999 was a staggering 89 million reptiles, a figure no doubt matched today as hundreds of thousands of hectares of the state's extraordinarily complex natural heritage is bulldozed, heaped and burnt. Apart from immediate individual casualties, the loss and fragmentation of habitat have implications at population and species levels. With the spectre of increased clearing for the expanding coal and gas exports and the push for more northern development, it is critical that habitat continuity be prioritised. And as pristine habitats decline, the role of regrowth vegetation as a repository for biodiversity cannot be overemphasised.

Over the past several years Queensland has endured record-breaking weather events. The state has lurched between floods, droughts and fires, including the catastrophic scorching of subtropical rainforests that have not burned for millennia. Temperatures are setting new maximum records annually. Scientists predict such patterns will become the 'new normal'. It seems certain that a changing climate has already begun to affect species and ecosystems, and these effects are likely to intensify. It is extraordinary that climate change remains as an item for debate rather than an accepted national and global peril requiring immediate action.

Other threats include feral cats, foxes, pigs, cane toads and fire ants.

The Condamine Earless Dragon *(Tympanocryptis condaminensis)* was widely recognised as an endangered species for several years prior its 2014 description. To garner money for research and conservation, local community groups raised its media profile and even sold chocolate effigies.

HOW TO USE THIS BOOK

Reptile identification can be complex. Many unnamed species lie in Australian museum collections and odd specimens still confound the experts. In a tidy world, features would be easy to see and invariable. The reader could accurately compare specimens with pictures and descriptions. Alas, microscopic examinations are sometimes necessary, though not usually possible in the field, and occasionally involve a degree of subjective interpretation. For example, the ear openings of Rainbow Skinks (*Carlia*), vital elements in identification, vary much more than is generally credited. Odd reptile specimens with missing or altered features may lead to a wrong answer.

With these limitations in mind, this book attempts to present reptiles in a format that most easily allows identification, using photographs, descriptions, identification keys, diagrams, maps and habitat notes. The reader should consider all, in tandem with other specialist publications, for the best results.

Photographs

Photographs represent typical examples rather than unusual colour forms. Unless captioned otherwise, specimens pictured are from Queensland.

Keys

Keys lead the reader through numbered couplets offering contradicting options to reach identification. Except those with only two readily distinguishable members, most families include keys to genera and most genera feature keys to species. Where appropriate, the features discussed are illustrated. If uncertain, the user can backtrack following the numbers in brackets. I have tried to base the keys on fixed and obvious characteristics, but they should be regarded as useful tools rather than the path to definitive results.

Descriptions

These are arranged in descending order of family, genus and species. Where a family or genus contains one species (monotypic), or has but a single species in Queensland, accounts are amalgamated. Features present on other species or populations outside Queensland are ignored. For example, the skink genus *Lerista* is listed as having 4 or fewer digits, although some species in southern States have 5 fingers and toes.

Generic text provides diagnostic features, distribution and distinguishing behavioural details applicable to all members. For example, Death Adders (*Acanthophis*) are cited as sedentary snakes that ambush prey from concealed sites beneath leaf litter, shared traits not repeated under each species.

Species accounts begin with a brief list of features considered most diagnostic. While colour and pattern are often unreliable in identification, those elements considered most typical are included. Following is a list of Queensland bioregions where the species occurs, short habitat notes and, where space permits, a note or two on natural history.

GLOSSARY AND ILLUSTRATIONS

These definitions apply to the context in which the words have been used in this book. In other contexts they can have different or additional meanings.

Adpress Pressing the forelimbs back and the hindlimbs forward along the sides of the body to assess relative limb length by whether they overlap and by how much.

Aestivation Inactivity during dry periods.

Auricular Relating to the ear.

Barbels Soft fleshy protuberances on the chins of some freshwater turtles.

Basal At or near the base.

Callose/callus A raised hard, tough or thickened condition, along scales under the digits.

Carapace Upper shell of a turtle.

Carination Keel or ridge on a scale.

Casque Helmet of thickened skin or bone over the head.

Caudal Relating to the tail.

Chevron Inverted V-shaped marking.

Cloaca Common chamber into which reproductive and excretory ducts open.

Crepuscular Active at dawn, dusk or in deeply shaded conditions.

Cryptic Inconspicuous or secretive by way of colour, pattern and/or behaviour.

Dewlap Loose flap of skin under the throat.

Distal Furthest from the body, away from the point of attachment.

Diurnal Active by day.

Dorsal Relating to the back or upper surfaces.

Dorsolateral Relating to the junction of the upper (dorsal) and side (lateral) surfaces. Usually refers to pattern.

Femoral pore One or more pores beneath the thigh.

Gular Relating to the throat, or in turtles, leading part of the plastron.

Heterogeneous Refers to that condition where the scales differ in size and/or shape.

Homogeneous Refers to that condition where the scales are similar.

Keel A narrow raised ridge on individual scales, or a low crest or other longitudinal flange.

Labial Of the lips, usually referring to scales bordering the lips.

Lamellae Scales along the underside of the digits.

Laterodorsal Outer part of the back, usually a dark stripe along the inner edge of a pale dorsolateral stripe.

Mental scale Anterior-most scale on the chin.

Monotypic Family or genus containing only one species.

Mucronate Ending in a sharp point or spine. Usually refers to acute subdigital lamellae of some lizards.
Nuchal Relating to the nape of the neck.
Ocelli Eye-like, ring-shaped spots.
Palpebral disc The transparent window in the lower eyelid of some lizards.
Paravertebral scales Longitudinal row of scales lying on each side of the mid-dorsal line.
Parthenogenetic Able to reproduce without fertilisation by a male, particularly some geckos that produce female clones.
Pectoral Part of the chest, or part of the plastron of a turtle.
Pelagic Inhabiting the open seas.
Plastron Lower part of a turtle shell.
Pore Opening to a duct in or between some scales on some lizards, best developed in males, usually filled with a wax-like substance.
Postauricular Behind the ear.
Postocular Behind the eye.
Preanal pore One or more pores located in front of the vent.
Preocular Immediately in front of the eye.
Proximal Nearest to the body, close to the point of attachment.
Reticulated Forming a net-like pattern or reticulum.
Rosette Circular arrangement of scales, usually surrounding a tubercle.
Rugose Rough.
Scansors Scale-like plates along the underside of the digits of some geckos. Setae are clustered on the scansors.
Scapula Dorsal part of the shoulder.
Scutes Enlarged scales on a reptile; horny plates of a turtle shell or bony plates (osteoderms) in crocodile skin.
Setae In relation to geckos, microscopic bristle-like structures with multiple-branched tips, clustered on the surface of the scansors, and used by application of Van Der Waal's Forces for molecular adhesion to smooth surfaces.
Spinifex Spiny-leafed grasses of the genera *Triodia* and *Plectrachne*, which form prickly hummocks in arid zones. Often called porcupine grass.
Striation Groove on the surface of a scale.
Subapical Beneath the tip, in reference to digits.
Subcaudal Beneath the tail. Often applies to scales.
Subdigital Beneath the finger or toe.
Suture The groove between non-overlapping scales.
Tubercle Rounded or pointed projection.
Vent Transverse external opening of the cloaca.
Ventral Of the lower surfaces, or the scales on the belly.
Ventrolateral Junction of the side (lateral) and lower surface (ventral).

1. Carapace of a turtle

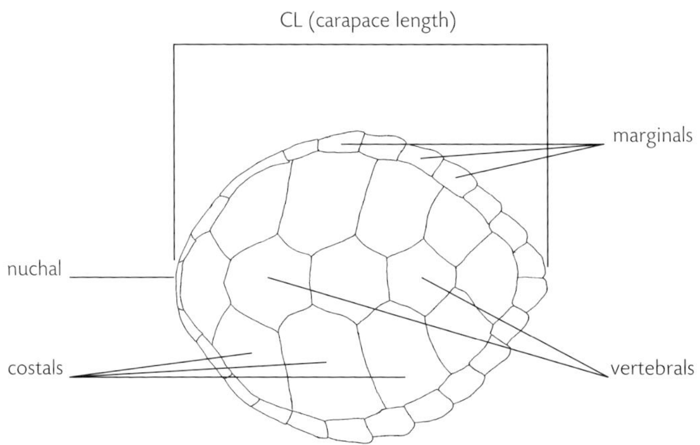

2. Reptile head shields (lateral)

3. Reptile head shields (dorsal)

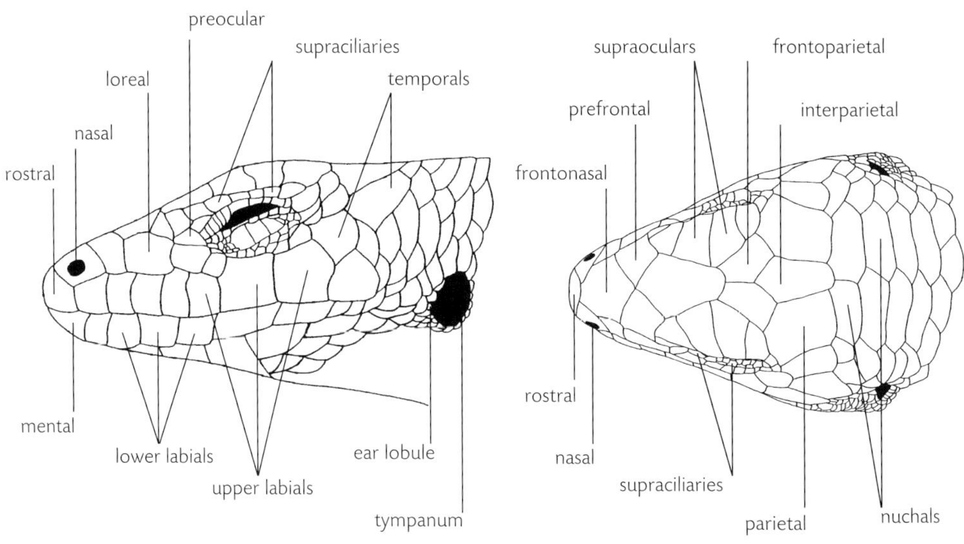

5. Midbody scales of a snake showing diagonal counts

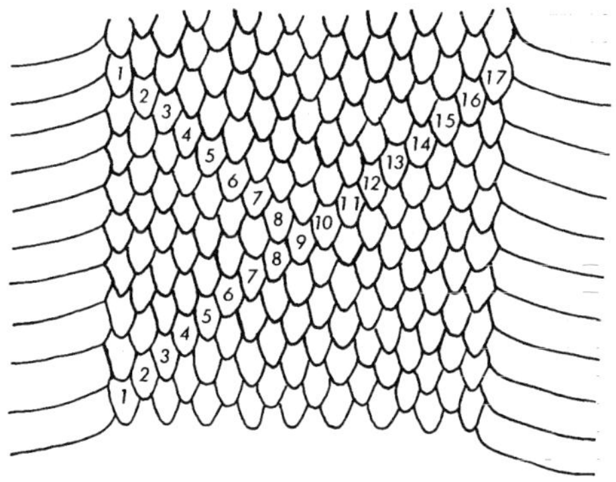

4. Lizard body indicating measuring points

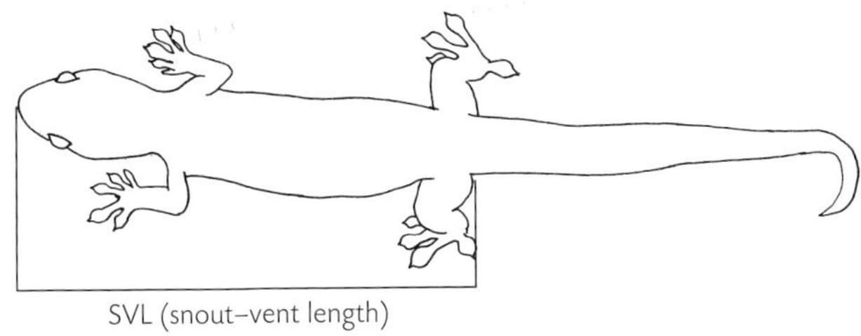

6. Variation in anal and subcaudal scales of a snake

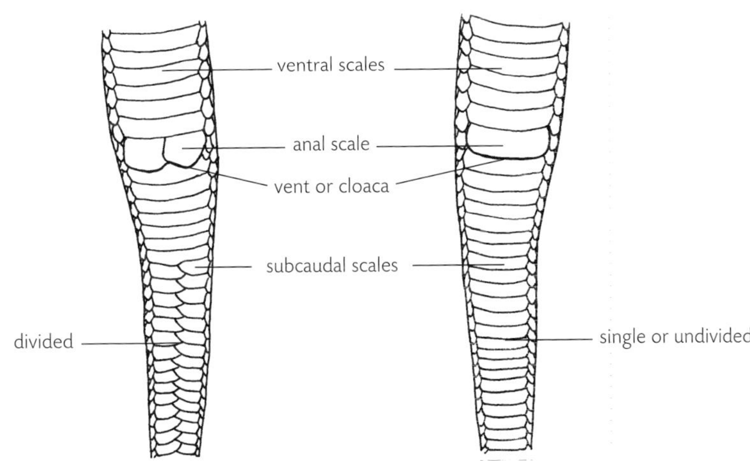

QUEENSLAND BIOREGIONS

Based on broad landscape patterns that reflect geology, climate and the distribution of floral and faunal communities, Queensland is divided into 13 bioregions. Each has distinctive characteristics, but their boundaries sometimes merge without sharp delineation. While they often include a number of quite separate and different habitats too complex to cover in detail, the bioregions form the basis for the distributional information in the following species accounts.

Northwest Highlands (NWH) Far northwest, bordering NT. Stony hills and ranges vegetated with eucalypt woodlands and spinifex. A number of river systems, some of which are spring-fed, flow perennially. Summer temperatures are high and rainfall is low and erratic.

Mitchell Grass Downs (MGD) A broad band of tussock grasslands (mainly *Astrebla* spp.) on deeply cracking clay across semi-arid centre and west. Largely featureless terrain, punctuated with isolated residual mesas with spinifex, and eucalypt-lined drainage systems.

Channel Country (CHC) Southwestern corner. Vast floodplains braided with innumerable tree-lined channels. A diverse region featuring clay pans, sand-plains and dunes with spinifex, and gibber plains and low weathered ranges with sparse low shrubs. Summer temperatures are extremely high and rainfall is low and erratic, with rare impressive floods heralding a burst of growth and activity.

Mulga Lands (ML) Semi-arid southern centre. Plains and low ranges dominated by Mulga (*Acacia aneura*) shrublands and woodlands on infertile red earth, mixing with eucalypts in the higher rainfall eastern parts. Large eucalypt-lined drainage systems, including the Paroo and Warrego Rivers, flow south into northwestern NSW.

Brigalow Belt (BB) Large and complex region (36 million hectares), encompassing a broad band from the NSW border to Townsville. It receives annual rainfall of about 500–750mm and features a wide range of landforms and vegetation communities, typically forests and woodlands of Brigalow (*Acacia harpophylla*), but also areas of eucalypts, dry rainforests and cypress pine.

New England Tableland (NET) Lower southeastern inland, adjacent to the NSW border. Cool temperate uplands, subject to regular winter frost and occasional snow, featuring massive granite outcropping, heath and dry sclerophyll forests. Some southern species reach their northern limits here.

Southeastern Queensland (SEQ) Coast and ranges between about Gladstone and the NSW border. Very diverse region with annual rainfall of 800–1500mm, up to

30 per cent of which falls in winter. Included are a coastal plain with heaths, paperbark woodlands and dry eucalypt forests, ranges with subtropical rainforests, mainland and island sand masses with heaths, woodlands and lowland rainforests, and the warm shallow waters of Moreton Bay.

Central Queensland Coast (CQC) Coastal lowlands and ranges in the Byfield/Shoalwater Bay area and from about Carmila to Proserpine. High rainfall (1300–2000mm per year) is seasonal, mostly between January and March. Lowlands feature woodlands and semi-deciduous rainforests while mountain ranges are dominated by dense rainforest and tall eucalypt forest.

Desert Uplands (DEU) Central north, straddling the Great Dividing Range between Blackall and Pentland. A semi-arid region with red to yellow infertile soils dominated by sandstone ranges and sand-plains.

Einasleigh Uplands (EIU) Northeastern interior, straddling the Great Dividing Range along the western edge of the Wet Tropics. Ranges and plateaus incorporating a variety of altitudes and climates, and diverse ecosystems. Mostly dry sclerophyll forests and woodlands, with pockets of dry rainforest in sheltered areas (e.g. the lava tubes of Undara).

Wet Tropics (WT) Northeast coast and ranges, from north of Townsville to south of Cooktown. Low coastal plains rising to rugged mountains (including Queensland's highest, Mt Bartle Frere, 1622m), with extensive plateau areas along the western margins. Annual rainfall ranges from over 8m to less than 1400mm. Extensively forested, from dense tropical lowland rainforest to mist-enshrouded upland heaths, and eucalypt forests in the drier areas. Contains a wealth of endemic species.

Cape York Peninsula (CYP) Complex and varied, including dune-fields and sand-plains with heaths, hills and gullies with rainforest, extensive eucalypt woodlands, dissected sandstone plateaus, continental islands of Torres Strait, and coral atolls and cays of the northern Great Barrier Reef. Rainfall is medium to high and strongly seasonal, with a winter dry season of at least seven months followed by monsoonal deluges.

Gulf Plains (GUP) Southern and eastern Gulf of Carpentaria, between the Mitchell River and the NT border. Mostly vast, grassy alluvial plains with open woodlands in depressions and along watercourses, extensive wetlands, coastal estuaries and mudflats. In the east it rises to gently sloping sandstone tablelands. Mainly dry winters and monsoonal summer rains, generally higher towards the coast and northeast.

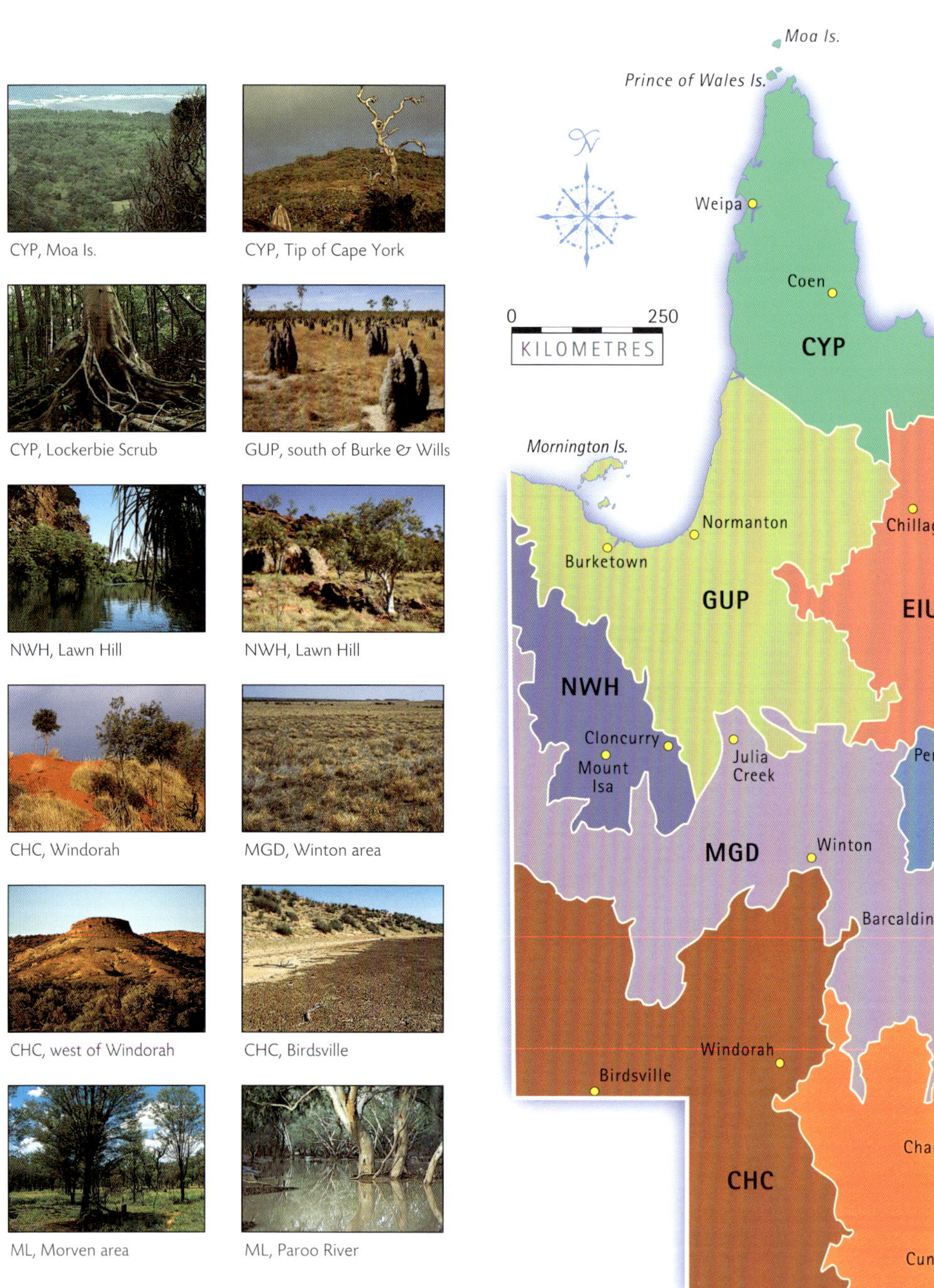

CYP, Moa Is.

CYP, Tip of Cape York

CYP, Lockerbie Scrub

GUP, south of Burke & Wills

NWH, Lawn Hill

NWH, Lawn Hill

CHC, Windorah

MGD, Winton area

CHC, west of Windorah

CHC, Birdsville

ML, Morven area

ML, Paroo River

KEY

CYP	Cape York Peninsula
GUP	Gulf Plains
EIU	Einasleigh Uplands
WT	Wet Tropics
NWH	Northwest Highlands
MGD	Mitchell Grass Downs
DEU	Desert Uplands
BB	Brigalow Belt
CQC	Central Queensland Coast
CHC	Channel Country
ML	Mulga Lands
NET	New England Tableland
SEQ	Southeastern Queensland

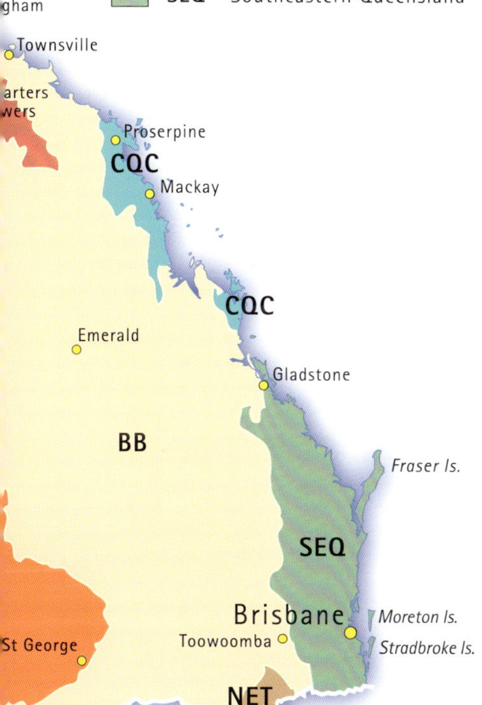

EIU, Forsayth

WT, Thornton Peak

EIU, Chillagoe

WT, Malanda

CQC, Mount Blackwood

CQC, Shoalwater Bay

DEU, Winhaven Station

SEQ, Fraser Island

BB, Westmar area

SEQ, Brisbane

BB, Yuleba area

SEQ, Mount Mee

NET, Girraween NP

SEQ, Maleny area

SNAKE BITE

Not all venomous snakes are dangerous. Most employ weak toxins sufficient to immobilise lizards and frogs, and have small fangs that are barely capable of puncturing human skin. Yet Australia's venomous snakes include some of the world's most lethal species, and even bites from mildly venomous animals can occasionally result in serious complications, so regard any bite or suspected bite as serious and seek immediate medical aid.

Effective treatment of snakebite can mean the difference between life and death. The 'pressure-immobilisation method' involves immobilising fluid within the lymph channels rather than arresting blood flow in the veins and arteries. The recommended First Aid for Australian snakebite is simple and effective.

- ✓ Keep the patient calm, lying down and still.
- ✓ Immediately apply a broad elastic bandage to the bitten area. Torn clothing or pantyhose will suffice if nothing else is available.
- ✓ Bind the entire limb, between toes and groin or fingers and armpit, as tightly as you would a sprain.
- ✓ Immobilise the limb with a splint and further bandaging.
- ✓ Clearly mark the site of the bite on the outside of the bandage.
- ✓ For bites to the trunk or face, maintain firm pressure to the site. Pressure-immobilisation cannot be applied to the head or neck.
- ✓ Take the patient immediately to medical aid.

DO NOT:

- ✗ **Wash the wound.**
- ✗ **Cut the wound.**
- ✗ **Apply a tourniquet to disrupt blood-flow in the limb.**
- ✗ **Attempt to kill or capture the snake.**

TEXT ABBREVIATIONS

adj. adjacent
approx. approximately
assoc. associated/association
Ck/s Creek/s
CL carapace length
e. east; estn eastern
excl. excluding
incl. including
Is. Island/s
juv. juvenile/s
Mtns Mountains
NP National Park
NG New Guinea
NZ New Zealand
n. north
ne. north-east
nthn northern
pop./pops population/s
Ra. Range/s
R. River
s. south
se. southeast
sthn southern
sp./spp. species (sing./pl.)
SF State Forest
Stn Station
ssp. subspecies (sing./pl.)
w. west
wstn western
SVL snout–vent length
TL total length
vegtn vegetation.

CROCODILES

Saltwater crocodiles (*Crocodylus porosus*) are the world's largest surviving reptiles. They are dangerous and have accounted for numerous fatalities. 'Salties' inhabit fresh, salt and brackish tropical waterways. Warning signs are erected at strategic points across northern Australia to dissuade anyone tempted to risk a refreshing dip. Daintree River.

Family **Crocodylidae**

Large reptiles with tough, bony dorsal scutes, short limbs with clawed webbed feet, long, laterally compressed tails, long snouts with nostrils set at the tip, and a notch on each side of the upper jaw accommodating the 4th tooth of the lower jaw. This tooth is always visible when the mouth is closed.

Crocodiles have changed little since the time of dinosaurs and winged reptiles. These aquatic predators in Australia's northern waterways hunt by stealth, mainly at night. Diets range from fish and frogs to large mammals – including people. By day crocodiles can often be seen basking on river banks. They are egglayers and guard their nests, sometimes assisting the young to hatch and transporting them to the water.

Once hunted to critically low levels, complete protection has resulted in a rise in numbers. Occasionally 'problem crocodiles' – large potential man-eaters – near populated areas must be trapped and removed.

Two species occur in Australia, both in the genus *Crocodylus.*

Freshwater Crocodile

Crocodylus johnstoni

TL 2–3m

Slender snout; single row of enlarged nuchal shields set fewer than 8 granular scale rows behind smooth skin at base of head; relatively small size. Grey to brown with regular darker bands.

Freshwater rivers, creeks and billabongs in NWH, GUP, EIU, CYP. Also nthn NT and WA. Feeds mainly on fish, captured with rapid sideswipe of slender snout, also crustaceans, insects, frogs. Rare injuries to people probably result from crocodile mistaking kicking foot for injured fish. Eggs laid in excavated burrow. Not considered dangerous.

Crocodylus johnstoni. Lake Moondarra.

S. Eipper

Estuarine Crocodile; Saltwater Crocodile

Crocodylus porosus

TL 3–5m (max. 7m)

Broad snout; 2 rows of enlarged nuchal shields set more than 8 granular scale rows back from smooth skin at base of head; extremely large size. Grey or brown to almost black with darker mottling.

Coastal rivers, swamps, estuaries and even open sea in GUP, CYP, WT, BB, CQC. Also nthn Aust. to wstn Pacific, Asia. World's largest crocodile. Thanks to size and broad snout, can dispatch more powerful prey than *C. johnstoni*. Eggs laid in mound of soil and composting vegtn. Responsible for numerous human fatalities here and overseas. DANGEROUS.

Crocodylus porosus. Karumba.

SEA TURTLES

Green Turtles (*Chelonia mydas*) occur along all of Queensland's coastline.

Families **Dermochelyidae** and **Cheloniidae**

These two families of wholly aquatic marine turtles have limbs modified to form flippers.

All sea turtles have a fully aquatic lifestyle, feeding, mating and basking in the water. Females haul themselves ashore only to bury their clutches of round white eggs; males need never leave the sea. They include herbivores that take sea grasses and algae, and carnivores that capture jellyfish, crustaceans, sponges and other invertebrates.

Feeding and breeding areas are normally separate, often hundreds or thousands of kilometres apart. Hard-shelled Sea Turtles (family Cheloniidae) are largely associated with tropical to subtropical continental shelves; most records from temperate waters represent lost strays. The Leathery Turtle (family Dermochelyidae) mainly dwells in cool temperate seas, often in open oceans well away from landmasses, but remains obligated to nest on tropical beaches.

All species of sea turtle are of conservation concern.

Family **Dermochelyidae**

The one member of this family in the genus *Dermochelys*, has enormous front flippers lacking claws and a tough, rubbery, ridged carapace with no separate scutes.

Leathery Turtle; Luth

Dermochelys coriacea

CL 1.8–2.8m

Long, streamlined, leathery carapace with prominent longitudinal ridges but no hard scutes; enormous front flippers with no claws; extremely large size (to 900kg). Very dark grey to black; little pattern except paler mottling. Dorsal ridge-lines of juv. beaded with white.

All coastal waters; SEQ records mainly winter. Also all States. Individuals roam global oceans to feed on jellyfish, mainly in cool temperate waters from s. of NZ to n. of Arctic Circle. Breeds in tropics, as few as 10 coming ashore annually in Aust. Limited nesting recorded between Fraser Is. and Mackay. Can dive to more than 1000m, and maintain body temperature higher than ambient water temperature. It is feared this long-lived, wide-ranging sp. is in severe decline.

C. Limpus

Dermochelys coriacea. Moore Park, Bundaberg area.

Hard-shelled Sea Turtles
Family **Cheloniidae**

The family Cheloniidae includes 5 species of hard-shelled sea turtles, each in a different genus. Front flippers are moderate-sized with one or more claws on the leading edge, and numerous tough scutes cover the shell (concealed beneath a veneer of thin skin on one species).

KEY TO *CHELONIIDAE*

1 Four costal scutes along each side of carapace 2
Five or more costal scutes along each side of carapace 4

2 One pair of large prefrontal scales **(a)**; non-overlapping scutes on carapace 3
Two pairs of prefrontal scales **(b)**; scutes on carapace overlapping .... ***Eretmochelys imbricata***

3 All scales large on upper surface of end half of front flipper **(c)**
.... ***Chelonia mydas***

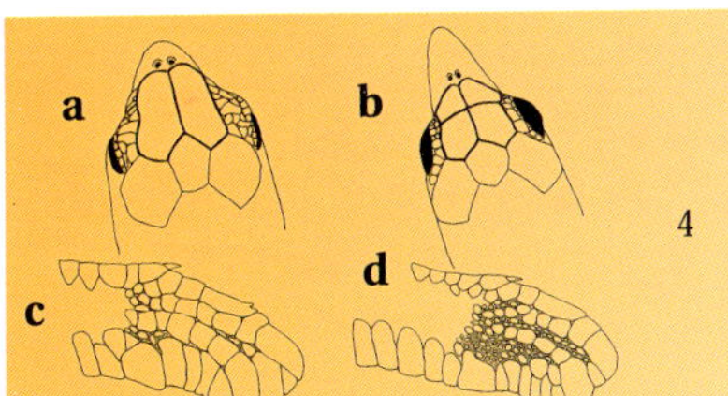

Rows of large scales on end half of front flipper, separated by minute scales or wrinkled skin **(d)** .... ***Natator depressus***

4 Five (rarely 6) costal scutes; carapace longer than wide; brown to reddish brown................................ ***Caretta caretta***
Six or more costal scutes; carapace roughly circular; olive grey................................ ***Lepidochelys olivacea***

Genus *CARETTA* Monotypic genus.

Loggerhead Turtle

Caretta caretta

CL 1.25m

Caretta caretta. Mon Repos.

Non-overlapping scutes on carapace; 5 (rarely 6) pairs costal shields; 2 pairs prefrontal scales; very large robust head and thick jaws, particularly on large adults. Shades of brown to reddish brown, sometimes with darker markings.

All coastal waters, with breeding records s. to SEQ. Most significant site Mon Repos near Bundaberg. Strays s. to Vic, SA. Mainly carnivorous, taking molluscs, crustaceans, sea urchins, jellyfish.

Genus *CHELONIA* Monotypic genus.

Green Turtle

Chelonia mydas

CL 1.5m

M. Shaw

Chelonia mydas. Raine Is.

Non-overlapping scutes on carapace; 4 pairs costal shields; 1 pair large prefrontal scales; series of large scales on upper eyelid. Olive green with black and darker brown markings, pale sutures highlighting head shields.

All coastal waters with breeding records s. to SEQ. Strays s. to Vic, SA. Adults largely herbivorous, feeding on sea grasses, algae, mangrove fruits; juv. carnivorous. Frequently eaten; common name derives from colour of its fat, not the animal.

Genus *ERETMOCHELYS* Monotypic genus.

Hawksbill Turtle

Eretmochelys imbricata

CL 1m

Overlapping scutes on carapace of adults (non-overlapping on juv.); 4 pairs costal shields; 2 pairs prefrontal scales; beak-like snout formed by jutting upper jaw; narrowly heart-shaped shell. Olive green to brown with prominent darker variegations, often forming radiating markings on each scute.

All coastal waters; strays reaching sthn NSW; no breeding s. of CYP. Adults omnivorous, taking mainly sponges but also molluscs, soft corals, sea grasses. Juv. carnivorous. Overlapping scutes basis of tortoiseshell industry, causing serious decline worldwide.

Eretmochelys imbricata. Milman Is.

C. Limpus

Genus *LEPIDOCHELYS* 1 Aust. species; 1 other in Gulf of Mexico.

Pacific or Olive Ridley Turtle

Lepidochelys olivacea

CL 60–75cm

Non-overlapping scutes on carapace; 6 or more pairs costal shields; 2 pairs prefrontal scales; nearly round shell. Olive grey with little pattern.

Waters of nthn CYP and WT. Also nthn NT. Single recorded stray in Vic. Nests in NT, Gulf of Carpentaria. Recorded to feed on urchins, small crabs, molluscs.

Lepidochelys olivacea. Cairns.

C. Limpus

Genus *NATATOR* Monotypic genus.

Flatback Turtle

Natator depressus

CL 80cm–1m

Thin veneer of skin over carapace of adults; very depressed carapace with upturned edges; non-overlapping scutes; 4 pairs costal shields; 1 pair large prefrontal scales; small scales over eye. Olive grey to pale green with little or no pattern.

All coastal waters s. to SEQ. Breeds as far s. as about Mon Repos; most significant breeding sites in north (e.g. Crab Is. off nw. CYP). Often comes ashore to nest by day. Only sea turtle to breed exclusively in Aust. Carnivorous, eating mainly soft-bodied invertebrates gathered from sea floor, jellyfish.

Natator depressus. Crab Is.

P. Couper

FRESHWATER TURTLES

A Worrell's Turtle (*Emydura worrelli*) surfaces for air among lily leaves in Lawn Hill NP.

Family **Chelidae**

Aquatic turtles with clawed, webbed feet and short to very long necks, withdrawn horizontally into the shell by folding sideways. The family occurs throughout most of Australia, excluding some desert regions, New Guinea and South America. They occupy all major freshwater drainage systems in Queensland, with diversity highest in permanent eastern waters. Those inhabiting temporary water bodies are often forced into aestivation, burrowing beneath the mud or well insulated surface debris during dry periods.

Freshwater turtles include a mix of carnivorous and herbivorous species, with most taking both animal and plant material. They feed in water but often emerge to bask on banks, protruding logs and rocks. At such times they tend to be extremely wary, plopping into the water at the slightest disturbance. The endemic Fitzroy River Turtle (*Rheodytes leukops*) rarely emerges or surfaces, drawing much of its oxygen directly from clear waters via its cloaca. Other species can emulate this to a limited degree.

Eggs are laid in burrows on land. Those of the Broad-shelled Snake-necked Turtle (*Chelodina expansa*) may take over a year to hatch.

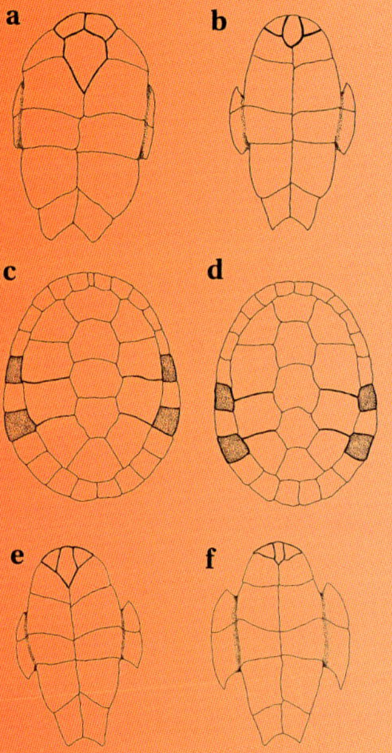

KEY TO GENERA

1	Gular shields in contact, completely enclosing intergular shield within plastron **(a)**	***Chelodina***
	Intergular shield contacts leading edge of plastron, separating gular shields **(b)**	2
2(1)	Sutures at rear of 2nd and 3rd costal scutes contact 6th and 8th marginal scutes **(c)**	***Rheodytes***
	Sutures at rear of 2nd and 3rd costal scutes contact 7th and 9th marginal scutes **(d)**	3
3(2)	Temporal region smooth; no horny casque on top of head	4
	Raised large scales or tubercles on temporal region; often a horny casque on top of head	5
4(3)	A pair of large barbels on chin; long tail at least 50 per cent of CL	***Elusor***
	Very small barbels on chin; tail much less than 50 per cent of CL	***Emydura***
5(3)	Intergular shield as wide as or wider than gular shields **(e)**; horny casque on top of head extends well down towards tympanum	***Wollumbinia***
	Intergular shield much narrower than gular shields **(f)**; horny casque not extending down towards tympanum	***Elseya***

Long-necked and Snake-necked Turtles

Genus *CHELODINA*

14+ spp. in Aust. and NG; 4 occur in Qld. Extremely long neck, usually as long as or longer than carapace; only 4 claws on forefeet; intergular shield does not contact leading edge of plastron.

Still and slow-flowing waterways of north, east, eastern interior. Also all mainland States. Ambush predators, lunging with the long neck while gaping mouth to suck in any creatures small enough to swallow. Infrequently seen basking but often encountered on land, moving between water bodies.

KEY TO *CHELODINA*

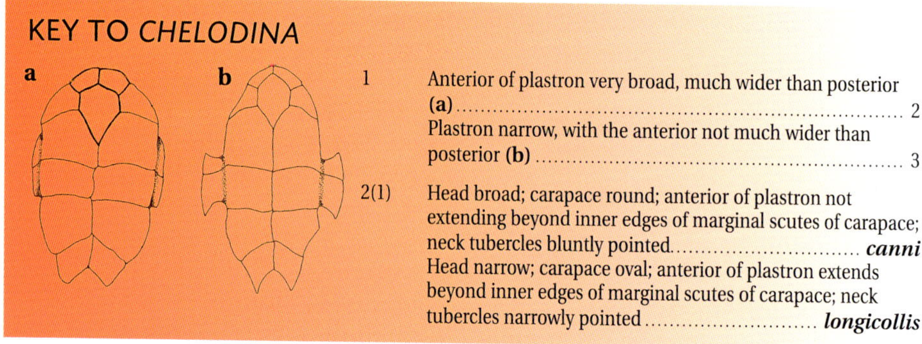

1	Anterior of plastron very broad, much wider than posterior **(a)**	2
	Plastron narrow, with the anterior not much wider than posterior **(b)**	3
2(1)	Head broad; carapace round; anterior of plastron not extending beyond inner edges of marginal scutes of carapace; neck tubercles bluntly pointed	***canni***
	Head narrow; carapace oval; anterior of plastron extends beyond inner edges of marginal scutes of carapace; neck tubercles narrowly pointed	***longicollis***

3(1) Plastron begins to taper immediately in front of bridge .... ***rugosa***

Plastron parallel-sided or slightly flared immediately in front of bridge .... ***expansa***

Cann's Long-necked Turtle

Chelodina canni

CL 264mm

Wide robust head; slender neck with bluntly pointed tubercles; wide round carapace, broadest at 7th marginal scute; anterior of plastron broader than posterior. Brown to almost black above with narrow pale rim to carapace. Yellow below with dark margins to scutes. Head, neck and limbs flushed with pink. Juv. very dark above, extensive red ventral pigment. Ill-defined ssp. *C. c. canni* and *C. c. rankini*.

Rivers, lagoons and swamps in GUP and CYP, also ne NT (*C. c. canni*); and WT, EIU and nthn BB (*C. c. rankini*).

A. Zimny

Chelodina canni canni. Archer River area.

Chelodina canni rankini. Ross River area.

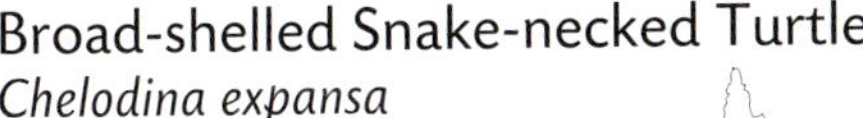

Broad-shelled Snake-necked Turtle

Chelodina expansa

CL 480mm

Extremely broad, strongly depressed head; very long, thick neck; very wide, flat carapace with expanded rear edges; narrow plastron; extremely large size. Brown to blackish brown, often with dark flecks or reticulations, whitish below.

Lagoons, lakes, rivers, swamps, incl. many broad inland drainage systems in SEQ, NET, BB, ML, CQC. Also interior of NSW, Vic, se. SA. Those from Fraser Is. very dark with narrower carapaces. Tends to favour silty water. Extremely secretive, lying concealed in debris on the bottom.

Chelodina expansa. Deception Bay, Qld.

Chelodina expansa. Deception Bay, Qld.

Eastern Long-necked Turtle
Chelodina longicollis
CL 238mm

Narrow head; slender neck with narrow pointed tubercles; moderately broad carapace, widest at 8th marginal scute; anterior of plastron much broader than posterior, extending beyond inner edges of marginal scutes. Shades of brown above, with dark sutures on pale animals. Paler brown to cream below, with sutures on plastron heavily outlined with black. Juv. boldly marked below with black and orange-red.

Occupies very wide range of wetland habitats, incl. swamps, lagoons, slow-moving rivers and creeks, and even isolated depressions that fill only briefly, in SEQ, NET, BB, ML, CQC, DEU. Also NSW, Vic, se. SA.

Chelodina longicollis. SE Qld.

Chelodina longicollis. Kogan area.

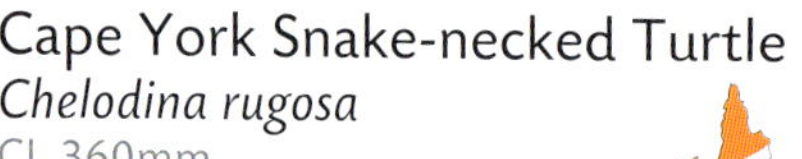

Cape York Snake-necked Turtle
Chelodina rugosa
CL 360mm

Broad, strongly depressed head; long thick neck; oval carapace, expanded posteriorly; narrow plastron. Dark brown to black above, pale below.

Large, slow-moving rivers, still waters of swamps and billabongs in CYP, EIU, GUP, NWH. Also nthn NT, WA. Frequently deposits eggs under water, where they remain dormant until level drops. During dry season along GUP, they bury themselves deep in the mud to await the rains.

J. Cann

Chelodina rugosa. Laradeenya Ck.

C. Dollery

Chelodina rugosa. Lakefield NP.

Snapping Turtles Genus *ELSEYA*

6 Aust. spp.; 4 occur in Qld, each occupying different river systems. Short neck, with head and neck shorter than carapace; 5 claws on each forefoot; intergular shield much narrower than gular shields, reaching leading edge of plastron to completely separate gular shields; sutures at rear of 2nd and 3rd costal scutes contact 7th and 9th marginal scutes; a horny shield over back of head not extending down toward tympanum; low rounded tubercles over temples. Juv. has a high central keel on carapace and serrated rear edge, normally becoming flatter with smoother edges with age.

Waterways of n. and ne. Aust. Also NG.

KEY TO *ELSEYA*

1	Head shield deeply grooved; Mary, Burnett and Fitzroy/Dawson drainages	***albagula***
	Head shield smooth ; drainage systems further north	2
2	Burdekin drainage system	***irwini***
	Drainage systems further north	3
3	North and South Johnstone River systems in Wet Tropics	***stirlingi***
	Nicholson drainage system in western Gulf of Carpentaria	***oneiros***

White-throated Snapping Turtle

Elseya albagula

CL 420mm

Among largest Aust. chelid turtles; deeply grooved head shield; rounded tubercles on neck; irregular pale markings on sides of head and neck; prominent serrated rear margin to carapace of juv. and young adults. Dark brown to black (adults) or tan (juv.); head and neck dark dorsally, and pale laterally and ventrally, the colours meeting along sharp irregular interface.

Endemic. Flowing waters in Fitzroy, Burnett and Mary Rivers and associated drainage systems of SEQ, CQC and BB. Primarily herbivorous, taking fallen fruit and buds of riparian vegetation and filamentous algae. Juv. more carnivorous.

J. Cann

Elseya albagula. Mary River.

Elseya albagula. Captive specimen.

Irwin's Snapping Turtle

Elseya irwini

CL 322mm

Very pale head; smooth head shield; low rounded and longer pointed tubercles on neck. Light brown with darker patches above, yellow piebald with black below. Head, incl. horny casque, very pale yellow, superficially appearing white. Snout, under eye to jaw, flushed rose pink.

Endemic. BB, EIU, estn DEU and WT, in clear waters of Burdekin, Johnstone and Hartley Creek drainage systems. Omnivorous. Plant material (incl. fallen riparian fruits) and snails recorded as food items.

G. Schmida

Elseya irwini. Broken River.

Gulf Snapping Turtle

Elseya oneiros

CL 320mm

Smooth head shield; suture between humeral and pectoral shields on plastron forms undulating line. Dark grey above, paler below, often with dense darker flecking on plastron, colours often masked by residue of plaque from mineral-rich waters. Feet, tail of males flushed reddish pink.

NWH. Restricted to Nicholson R. drainage of Qld and NT.

Elseya oneiros. Lawn Hill NP.

Stirling's Snapping Turtle

Elseya stirlingi

CL 340mm

Large robust head with smooth head shield; medium-sized tubercles on neck. Black above and bone to pale yellow below. Head and neck grey above and cream below, often with pale piebald pattern on mature females.

Elseya stirlingi. Malanda Falls.

D. Trembath

Elseya stirlingi. Johnstone River.

Endemic. North and South Johnstone Rivers, WT, with adults occupying deep pools and juveniles often in shallow water riffle zones.

Genus *ELUSOR* Monotypic genus.

Mary River Turtle

Elusor macrurus

CL 400mm

Head and neck shorter than carapace; long sharp neck tubercles; 2 very large barbels on chin; extremely long thick tail; 5 claws on each forefoot; intergular shield completely separates gular shields; carapace oval to slightly wider posteriorly, with central keel and small serrations along rear edges on juv., becoming flat with smooth edges on adults; sutures at rear of 2nd and 3rd costal scutes contact 7th and 9th marginal scutes. Shades of brown above, grey below.

Endemic. Restricted to Mary R. drainage, SEQ. Omnivorous.

Elusor macrurus. Mary River, Kenilworth area.

Genus *EMYDURA*

7 Aust. spp.; 5 occur in Qld. Head and neck shorter than carapace; neck smooth or with low rounded tubercles; small barbels on chin; 5 claws on each forefoot; intergular shield completely separates gular shields; carapace with weak to strong central keel and serrations along rear edges on juveniles, becoming flatter with smoother edges on adults; sutures at rear of 2nd and 3rd costal scutes contact 7th and 9th marginal scutes; pale streak on face

Still waters throughout Qld, incl. interior drainages. Also nthn and estn Aust., and NG. Omnivorous, consuming aquatic plants, fallen fruits, carrion, invertebrates and small vertebrates. Macrocephaly common in nthn populations. They often bask on protruding rocks or logs.

KEY TO *EMYDURA*

1	Obvious pale streak from behind eye to ear or beyond on all but aged specimens	2
	Pale streak either absent or narrow and not reaching ear	5

2(1)	Yellow pale streak behind eye	3
	Pink, orange to bright yellow pale streak behind eye	4
3(2)	Dark horizontal streak through pupil	***tanybaraga***
	No dark streak through pupil	***krefftii krefftii***
4(2)	Jardine River system, far northern Cape York	***subglobosa***
	Drainages of western Gulf of Carpentaria	***worrelli***
5(1)	Fraser Island only	***krefftii nigra***
	Excluding Fraser Island	6
6(5)	South-east Queensland and Murray/Darling drainages	***macquarii***
	Cooper Creek drainages	***krefftii emmotti***

Krefft's Turtle; Cooper Creek Turtle; Fraser Island Short-necked Turtle

Emydura krefftii

CL 250–368mm

Pale yellow streak from corner of jaw along neck. Krefft's Turtle (*E. k. krefftii*) is pale brown to dark brown above, with yellow streak behind eye. Iris variable but typically yellow around pupil fading to green on outer rim. CL 342mm. Cooper Creek Turtle (*E. k. emmotti*) is largest (CL 368mm) and very deep-bodied, with patternless face on adults and green iris with a bright yellow inner ring. Fraser Island Short-necked Turtle (*E. k. nigra*) is smallest (250mm) with dark skin and carapace and green iris with a yellow inner ring. *E. k. krefftii* is said to be the only turtle on the eastern seaboard with a pale facial streak, though it is sometimes present on other species.

East-flowing drainages from just n. of Brisbane to Princess Charlotte Bay, encompassing SEQ, BB, CQC, EIU, WT and CYP (*E. k. krefftii*); rivers, creeks and lagoons of the Cooper Ck drainage in CHC and MGD (*E. k. emmotti*); lakes of Fraser Island (*E. k. nigra*)

Emydura krefftii krefftii. Burnett River.

Emydura krefftii krefftii. Mary River, Kenilworth.

J. Cann

Emydura krefftii nigra. Fraser Is.

A. Emmott

Emydura krefftii emmotti. Lochern NP.

J. Cann

Emydura krefftii emmotti. Cooper Creek.

Macquarie Turtle; Brisbane River Turtle
Emydura macquarii
CL 275–340mm

Pale yellow streak from corner of jaw along neck. Pale brown to dark brown above, normally with little or no pale marking behind eye. Edge of carapace turns up at marginal scutes 4–7. Macquarie Turtle (*E. m. macquarii*) male has distinctly broad rear carapace and relatively narrow anterior plastron. Female is moderately large (340mm). Brisbane River Turtle (*E. m. signata*) male has more oval carapace, relatively broad anterior plastron. Female is smaller (CL 275mm).

Rivers, creeks and lagoons of sthn and wstn flowing drainages of the Murray/Darling Basin in BB, NET and ML, extending into NSW, Vic and SA (*E. m. macquarii*), and Brisbane River drainages in SEQ and ne NSW (*E. m. signata*).

Emydura macquarii macquarii. Stanthorpe district.

Emydura macquarii signata. Sandgate.

Jardine River Turtle; Painted Turtle
Emydura subglobosa
CL 250mm

Very prominent orange to pink stripe from snout, through top of eye to above ear, and from mouth along side of neck; dark horizontal streak through eye. Facial markings become weak to absent on aged macrocephalic specimens. Pink to red on limbs, plastron and lower edges of carapace.

Emydura subglobosa. NG.

Jardine R. drainage on far nthn CYP, n. and NG.

Northern Yellow-faced Turtle
Emydura tanybaraga
CL 285mm

Prominent yellow stripes from eye to above ear, and along side of neck; dark horizontal streak through eye. Brown to dark brownish grey with scattered dark spots above. Facial pattern fades on large adults but distinctive iris pattern always remains.

Mitchell R. and associated drainages of GUP, EIU, wstn CYP, approaching wstn edge of WT near Mossman. Also nthn NT.

Worrell's Turtle
Emydura worrelli
CL 260mm

Very prominent orange to pink stripe from snout, through top of eye to above ear, and from mouth along side of neck. Facial markings weaker on aged specimens. Carapace brown and plastron cream.

Waterways along wstn Gulf of Carpentaria, e. to Saxby R, NWH, GUP.

J. Cann

Emydura tanybaraga. Mitchell River.

Emydura worrelli. Lawn Hill NP.

Genus *RHEODYTES* Monotypic genus.

Fitzroy River Turtle
Rheodytes leukops
CL 250mm

Extended head and neck much shorter than carapace; large pointed tubercles on neck; broadly oval carapace (rounder with strong serrations along rear edge on juv.) with sutures at rear of 2nd and 3rd costal scutes contacting 6th and 8th marginal scutes; 5 claws on each forelimb. Shades of brown above. Pupil grey with white ring around iris. Large males have prominent orange blotches on sides of neck and throat.

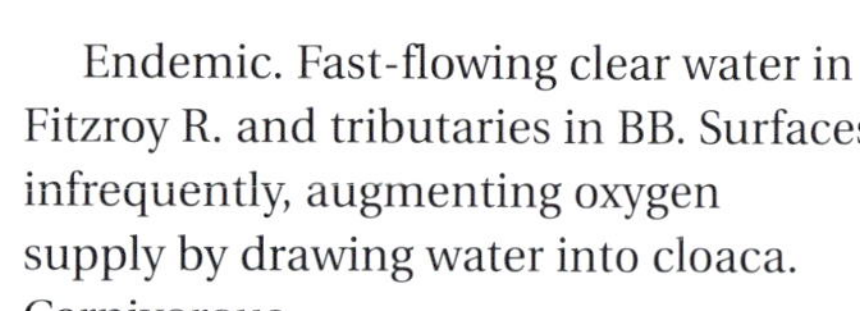

Endemic. Fast-flowing clear water in Fitzroy R. and tributaries in BB. Surfaces infrequently, augmenting oxygen supply by drawing water into cloaca. Carnivorous.

J. Cann

Rheodytes leukops. Fitzroy River.

Helmeted Turtles
Genus *WOLLUMBINIA*

3 Aust. spp.; 2 occur in Qld. Short neck, with head and neck shorter than carapace; 5 claws on each forefoot; intergular shield as broad as or broader than gular shields, reaching leading edge of plastron to completely separate gular shields; sutures at rear of 2nd and 3rd costal scutes contact 7th and 9th marginal scutes; a horny shield over back of head extending downwards and forwards toward tympanum; sharp-pointed tubercles on neck. Juv. and young adults have a high central keel on carapace and a serrated rear edge, normally becoming flatter with smoother edges with age.

Waterways of n. and ne. Aust. Omnivorous to primarily carnivorous.

KEY TO *WOLLUMBINIA*

1 West-flowing upland drainage, Bald Rock Ck, NET; iris clear .... ***belli***

2 Widespread but restricted to east-flowing drainages in s.; dark spots in iris, anterior and posterior to pupil .... ***latisternum***

Bald Rock Creek Turtle
Wollumbinia bellii
CL 290mm

An indentation in centre of each rear marginal scute forming serrations. Dark grey to black above and dark grey below, with steely grey head and neck; and yellow stripe along neck, darkening with age. Mature turtles have black plastrons.

Known only from Bald Rock Ck, a w. flowing system at about 800 m altitude in NET. Affinities with true *W. bellii* of the w. flowing MacDonald, Namoi, and Gwydir River systems of ne. NSW are uncertain and the Bald Rock Creek pop. may be an endemic, very restricted sp.

Wollumbinia bellii. Bald Rock Ck, Girraween NP.

Wollumbinia bellii. Bald Rock Ck, Girraween NP.

Saw-shelled Turtle
Wollumbinia latisternum
CL 287mm

Serrations along rear of carapace are prominent in young, and range from weak to prominent in adults from different locations; often dark spots in iris, anterior and posterior to pupil. Brown to black above and cream or yellow to dark grey below. A pale streak sometimes present along neck.

Lagoons, creeks and upper reaches of larger rivers of e. flowing drainages in SEQ, BB, CQC, DEU, EIU and WT and estn CYP, then w. to Gregory and Nicholson catchments in GUP and NWH. Also NT and NSW. Feeds on plant and animal material, including occasional cane toads.

Wollumbinia latisternum. Lake Kurwongbah.

Wollumbinia latisternum. Large adult with macrocephaly. Lake Kurwongbah.

Slider Turtles
Family **Emydidae**

One introduced aquatic turtle in the large genus *Trachemys*, with clawed, webbed feet and short neck, withdrawn directly back into shell. The family occurs in North America, Europe and parts of South America.

Red-eared Slider
Trachemys scripta elegans
CL 280mm

Direct retraction of head into shell (all native spp. withdraw head sideways); bright colouration on body and shell; extremely long claws on forefeet of males. Head and neck with numerous very narrow pale stripes, broad red stripe behind each eye, yellow bars and narrow lines on carapace, dark blotch on each plastron scute.

Breeding pop. exists in creeks and ponds at Mango Hill/Narangbah area on Brisbane's nthn outskirts. Introduced via pet trade from North America. Young carnivorous; adults mainly herbivorous. Can deliver painful bite when handled.

Trachemys scripta elegans. Mango Hill.

CHAMELEON, THICK-TAILED AND KNOB-TAILED GECKOS

A Granite Belt Thick-tailed Gecko (*Uvidicolus sphyrurus*) rears and lunges at the camera, demonstrating the spectacular threat display characteristic of many carphodactylid geckos. Girraween NP.

Family **Carphodactylidae**

These geckos represent the only reptile family endemic to Australia. There are 32 described species, with 20 in Queensland. They share important characteristics with the flap-footed lizards (Pygopodidae) and two other Australian gecko families, Gekkonidae and Diplodactylidae.

Immoveable transparent spectacles (brilles) cover the eyes; eye surfaces are cleaned with their broad flat tongues; and they lay just two eggs per clutch. These are soft and parchment-like. When harassed or during social encounters Carphodactylid geckos raise their bodies, wave their tails and utter a rasping squeak or bark.

Most are large species with flattened or box-shaped heads. They have slender limbs lacking any pads on the digits. Most have highly modified, sometimes spectacular tails, ranging from leaf- to carrot-shaped. These are easily discarded but have just one cleavage point at the base. Tails lost are forfeited completely.

KEY TO GENERA

1 Body strongly laterally compressed including raised vertebral ridge .... ***Carphodactylus***
Body moderate to dorsally flattened 2

2(1) Original tail terminates in a small knob; no enlarged lamellae under digits .... ***Nephrurus***
No knob on original tail; enlarged lamellae under digits 3

3(2) Tail swollen 4
Tail extremely broad and flat to long, tapering and cylindrical 5

4(3) Anterior loreal scales minute and strongly differentiated from larger posterior loreal scales .... ***Underwoodisaurus***
Anterior loreal scales only slightly smaller than posterior scales .... ***Uvidicolus***

5(3) Nostril separated from rostral scale **(a)** .... ***Phyllurus***
Nostril contacting rostral scale **(b)** 6

6(5) Extremely long neck; McIlwraith Range on CYP only .... ***Orraya***
Moderate neck; Cape Melville on CYP southwards .... ***Saltuarius***

Genus *CARPHODACTYLUS* Monotypic genus.

Chameleon Gecko
Carphodactylus laevis
SVL 130mm

Strongly laterally compressed body; uniformly fine scales with no enlarged tubercles; large, forward-directed eyes; long slender limbs; slender padless digits; carrot-shaped original tail with sharp black and white rings; simple regenerated tail with drab mottling. Brown with narrow pale vertebral line. Snout flushed cream above, dark on sides.

Endemic. WT. Rainforests between Cardwell Ra., Mt Windsor Tableland. Forages amongst leaf litter or perches head downward on slender stems, probably scanning for prey. Tail readily discarded; when regenerated tail is severed it produces loud squeaking noise as it wriggles.

Carphodactylus laevis. Wooroonooran NP.

Knob-tailed Geckos
Genus *NEPHRURUS*

10 Aust. spp.; 2 occur in Qld. Unique, small, round knob on tip of original tails; original tails fat and heart-shaped to greatly reduced; large, deep heads; plump bodies; slender limbs with short, clawed, padless digits, lacking any enlarged subdigital lamellae; fine granular body scales mixed with simple small scattered tubercles, or with prominent rosettes comprised of large conical tubercles surrounded by smaller tubercles.

Sandy deserts to dry tropical woodlands. These terrestrial geckos excavate burrows or commandeer those of other animals, emerging at night to forage for arthropods and smaller geckos. If provoked, they raise their bodies high on slender limbs like angry cats, lunging with mouth agape and uttering a loud rasping squeak.

Prickly Knob-tailed Gecko
Nephrurus asper
SVL 115mm

Tail greatly reduced, though spherical knob remains prominent; body and limbs conspicuously covered with spiny rosettes. Brown with bands of paler spots alternating with fine dark transverse lines; fine dark network over head. Far nthn populations prominently marked with dark and pale bands of about equal width.

Endemic. Mainly dry woodlands and rocky hills in BB, MGD, CHC, NWH, DEU, EIU; also sandy heaths in nthn CYP. One of 3 sibling spp. (others in NT, WA) with tails so reduced that ability to discard them appears lost; probably unique among geckos.

Nephrurus asper. Capella.

Smooth Knob-tailed Gecko
Nephrurus levis
SVL 102mm

Tail plump, heart-shaped; scales fine, mixed with small simple tubercles. Pale pinkish brown with narrow pale bands across base of head, neck and shoulders.

Arid sandy zones in CHC, ML. Most common on dunes with spinifex. Encountering these geckos patrolling the cooling desert sands after sunset, it is difficult to fathom how such delicate-looking, soft-bodied creatures can thrive in burrows a few cm underground when surface temperatures reach lethal extremes daily. Represented in Qld, all mainland States except Vic, by ssp. *N. l. levis*. Other ssp. in WA.

Nephrurus levis. Windorah.

Genus *ORRAYA* Monotypic genus.

McIlwraith Leaf-tailed Gecko
Orraya occulta
SVL 108mm

Broad flat tail, original teardrop-shaped with margin of clustered spines, regenerated smoothly round with spiny edges; extremely long neck; strongly depressed head and body; long slender limbs with angular, clawed, padless digits; rostral scale contacts nostril; small granular scales mixed with larger conical tubercles over body, limbs and original tail. Grey with simple dark mottling.

Endemic. Restricted to granite boulders in rainforest along Peach Ck, McIlwraith Ra., CYP. Shelters by day in cavities between boulders, at night rests motionless on exposed rock faces.

Orraya occulta. McIlwraith Ra.

Genus *PHYLLURUS*

10 Aust. spp.; 9 occur in Qld. Original tails cylindrical and simple, to extremely flat with plain flared rim and cylindrical pointed tip, all with conical tubercles; regenerated tails usually have shorter tips, more elaborate flares, no conical tubercles; strongly depressed head and body; long slender limbs with angular, clawed, padless digits; rostral scale separated from nostril; small granular scales mixed with larger conical tubercles over body, limbs and original tail. Pattern comprises simple, lichen-like flecks, pale bands on original tails and plain mottling or flecking on regenerated tails.

Rainforests and vine thickets along e. coast and ranges. None occur together, though near Mackay narrow distributions of 4 spp. virtually abut, separated only by seemingly insignificant bands of dry lowlands between discrete rainforest enclaves. They hide by day in rock crevices and tree hollows, emerging at night to rest motionless, invisible and usually head downward.

KEY TO *PHYLLURUS*

a

b

1	Tail cylindrical	2
	Tail broadly to narrowly leaf-shaped	5
2(1)	Rostral scale completely divided by a midline groove (**a**)	4
	Rostral scale partly divided (**b**)	3
3(2)	Internasal scales 5–8; subdigital lamellae under 4th toe 16–19	***gulbaru***
	Internasal scales 9–10; subdigital lamellae under 4th toe 19–21	***pinnaclensis***

4(2) Ventral surface of hind limb with small granules intermixed with pronounced raised tubercles........ ***caudiannulatus***
Ventral surface of hind limb with uniform granules........ ***kabikabi***

5(1) Rostral scale completely divided by one groove........ ***championae***
Rostral partly divided by one or more grooves........ 6

6(5) Ventral surface with dark peppering........ ***nepthys***
Ventral surface plain........ 7

7(6) Anterior pale band on original tail unbroken; body tubercles large........ 8
Anterior pale band on original tail broken on midline; body tubercles very small........ 10

8(7) Throat scales interspersed with larger tubercles........ ***ossa tamoya***
Throat scales uniformly small and granular........ 9

9(8) Large, sharply pointed dorsal tubercles on lower forearm extend to wrist........ ***ossa ossa***
Dorsal tubercles decrease in size and spinosity on lower forearm above wrist........ ***ossa hobsoni***

10(7) Original tail mainly black; ventral surface of original and regenerated tails dark with pale spots and blotches........ ***isis***
Original tail mainly brownish grey; ventral surface of original and regenerated tails pale with dark spots and blotches........ ***amnicola***

Mt Elliot Leaf-tailed Gecko

Phyllurus amnicola

SVL 113mm

Tail flat and flared, attenuate tip terminating in minute knob; small tubercles on body; rostral scale partly divided by deep vertical groove. Beige with irregular darker blotches on head, body and limbs. Original tail dark greyish brown with pale bands, 1st broken on midline.

Endemic. Recorded from granite boulders in rainforest along a creekline at 400–1000m on Mt Elliot, Bowling Green Bay NP, nthn BB.

Phyllurus amnicola. Regenerated tail. Mt Elliot.

Ringed Thin-tailed Gecko

Phyllurus caudiannulatus

SVL 103mm

Tail narrow and cylindrical; large tubercles on body and tail; pronounced raised tubercles beneath hind limb; rostral scale completely divided by a midline groove. Grey to brown with fine dark mottling or blotches, pale spots, and 5–6 pale bands on original tail.

Phyllurus caudiannulatus. Bulburin NP.

Endemic. Known only from RF in Bulburin NP, SEQ. The most favoured sites are large hollow figs which support large numbers within their labrynthine cavities.

Champion's Leaf-tailed Gecko

Phyllurus championae

SVL 80mm

Tail flat and flared, to carrot-shaped, attenuate tip terminating in minute knob; tubercles small on back and large on flanks; rostral scale completely divided (very rarely partly divided). Brown with irregular dark blotches over head, body, limbs. Original tail brown with heavy black mottling and 5 white bands.

Endemic. Narrowly restricted to 2 rainforest blocks, Cameron Ck and Blue Mtn, only 21km apart, in CQC. Most found at night on rocks, or on tree trunks near rocks.

Phyllurus championae. Cameron Creek.

Gulbaru Leaf-tailed Gecko

Phyllurus gulbaru

SVL 93mm

Tail narrow and cylindrical; small tubercles on body, larger tubercles on base and along sides of original tail; 5–8 internasal scales in a line above rostral scale; 16–19 subdigital lamellae under 4th toe; rostral scale partly divided by midline groove. Grey with irregular dark blotches on head, body and limbs, pairs of larger dark and pale blotches on hips, and 8 pale bands on dark grey original tail.

Endemic. Interface between far nthn BB, sthn WT. Known only from steep gullies with boulders, hoop pine and rainforest at Hervey Ra. and sthn end sthn end of Paluma Ra.

Phyllurus gulbaru. Hervey Ra.

Mt Blackwood Broad-tailed Gecko

Phyllurus isis

SVL 76mm

Tail flat and flared, long attenuate tip terminating in minute knob; tubercles greatly reduced, indistinct on back and small on flanks; rostral scale partly divided by groove; small size. Greyish brown with dark blotches over head, body and limbs. Original tail black with dense pale mottling, 2 pale bands broken by dark vertebral stripe on flared portion, and 3 bands on slender tip. Regenerated tail greyish brown with irregular pale blotches.

Endemic. Known only from sheltered rocky outcrops on 2 small rainforest peaks, Mt Blackwood and Mt Jukes, just n. of Mackay, CQC.

Phyllurus isis. Mt Blackwood.

Oakview Leaf-tailed Gecko

Phyllurus kabikabi

SVL 81mm

Tail narrow and cylindrical; large tubercles on body and tail; no raised pronounced tubercles beneath hind limb; rostral scale completely divided by a midline groove. Grey to brown with fine dark mottling or blotches, obscure pale spots on either side of vertebral line and 5 pale bands on original tail.

Endemic. Known only from narrow band of semi-evergreen vine thicket on fragmented rocks capping a ridge at 540 metres elevation in Oakview NP near Kilkivan, SEQ.

Phyllurus kabikabi. Oakview NP.

Eungella Broad-tailed Gecko

Phyllurus nepthys

SVL 103mm

Tail flat and flared, to carrot-shaped; large, very prominent conical tubercles over body, limbs and tail; rostral scale partly divided by single groove; peppered ventral surfaces. Shades of brown with fine darker and paler blotches. Original tail brown with dark blotches, 4–5 prominent pale bands. Regenerated tail simply mottled with cream and dark brown.

Phyllurus nepthys. Eungella NP.

Endemic. Confined to rainforest in Eungella NP and surrounding areas of Clarke Ra., CQC.

Mount Ossa Leaf-tailed Gecko

Phyllurus ossa

SVL 89mm

Tail flat and flared, to carrot-shaped, attenuate tip terminating in minute knob; conical tubercles of moderate size over body, limbs and tail; rostral scale variable, its top usually notched by 3 deep grooves, sometimes by 2 or rarely an inverted 'Y-shaped' groove. Grey-brown with darker blotches and paler spots over head, body, limbs and flared portion of original tail. Remaining slender portion black. Original tail banded with cream; regenerated tail with dark and pale blotches. Throat usually smooth, but *P. o. tamoya* has scattered enlarged

Phyllurus ossa hobsoni. Mt Dryander.

Phyllurus ossa tamoya with regenerated tail. Whitsunday Island.

tubercles. *P. o. ossa* has uniformly large tubercles on forearm; on *P. o. hobsoni* these become smaller towards wrist.

Endemic. Separate pops occur in rocky areas within RF in CQC: Mt. Ossa/ Mt. Charleton/Mirani area near Mackay (*P. o. ossa*); Conway Range/Mt. Dryander area near Proserpine (*P. o. hobsoni*); and Whitsunday Is. (*P. o. tamoya*).

Phyllurus ossa ossa. Mt Ossa.

Pinnacles Leaf-tailed Gecko

Phyllurus pinnaclensis

SVL 93mm

Tail narrow and cylindrical; small dorsal tubercles on body, larger on flanks and mainly absent from basal quarter of tail; 7–10 (usually 9–10) internasal scales in a line above rostral scale; 19–21 subdigital lamellae under 4th toe; rostral scale partly divided by a midline groove. Pale grey with darker blotches and 5 prominent pale bands across original tail.

Endemic. Restricted to vine forest on deeply fissured rock at The Pinnacles near Townsville, nthn BB.

Phyllurus pinnaclensis. The Pinnacles. C. Hoskin

Genus *SALTUARIUS*

7 Aust. spp.; 5 occur in Qld. Original tails very broad, flat and elaborately flanged with spiny cylindrical slender tip (smooth with short blunt tip when regenerated); strongly depressed head and body; long slender limbs with angular, clawed, padless digits; small granular scales mixed with larger conical tubercles over body, limbs and original tail; rostral scale contacts nostril. Bold, complex blotches and streaks combine to break the outline and create lichen-like marbling. Largest and most impressive leaf-tailed geckos.

Rainforests and major granite and sandstone outcrops from CYP to NET. Also nthn NSW. By day they shelter on cave walls, in rock crevices and hollow tree trunks. At night they emerge to rest head downward, waiting to ambush passing prey.

KEY TO *SALTUARIUS*

a b c d

1 Eye unpatterned; elaborately frilled margin on original tail; Cape Melville on CYP only .... ***eximius***
Eye with blotched iris; margin of original tail with spines but no frill; WT southwards 2

2(1) Throat smooth 3
Throat rough .... ***salebrosus***

3(2) Flank tubercles conical, surrounded by granular scales (**a**).. 4
Flank tubercles hooked, surrounded by smaller spines (**b**).... .... ***cornutus***

4(3) Large tubercles on thick-tipped original tail (**c**) .... ***swaini***
Small tubercles on fine-tipped original tail (**d**) ...... ***wyberba***

Northern Leaf-tailed Gecko
Saltuarius cornutus
SVL 144mm

Long, backward-curved tubercles surrounded by smaller spines on flanks; smooth throat; attenuated tip of original tail more than one-third total tail length; very large size.

Endemic. WT, between Cooktown and sthn end of Paluma Ra. Shelters by day in cavities within trunks of rainforest trees. Normally seen at night resting on trunk of home tree, on the ground or nearby sapling.

Saltuarius cornutus. Woonooroonan NP.

Cape Melville Leaf-tailed Gecko
Saltuarius eximius
SVL 120

Extremely slender with long limbs, short head, extremely large unpatterned eyes; original tail with elaborate, undulating frilled margin and fine attenuate tip less than one third total tail length. Blackish brown to black with bold irregular pale grey bands.

Endemic. Isolated RF uplands above granite boulders on Melville Ra. in Cape Melville NP, CYP.

C. Hoskin

Saltuarius eximius. Cape Melville.

Rough-throated Leaf-tailed Gecko
Saltuarius salebrosus
SVL 140mm

Short conical tubercles on flanks; prominent tubercles on throat; attenuated tip of original tail less than one-third total tail length; very large size.

Saltuarius salebrosus. Bulburin SF.

Endemic. BB, CQC, far nthn SEQ, from Blackdown Tableland s. to Cracow and inland to Carnarvon Ra. Rainforest pops occur at Bania and Bulburin SFs; rock-inhabiting pops occupy granite and sandstone outcrops, cliffs and caves.

Saltuarius swaini. Springbrook.

Southern Leaf-tailed Gecko

Saltuarius swaini

SVL 131mm

Short conical tubercles on flanks; smooth throat; attenuated tip of original tail more than one-third total tail length, relatively thick with large tubercles; narrow, deep V-shaped mark between eyes; moderate size.

Rainforests of SEQ, from Border Ra. n. to Mt Glorious. Also ne. NSW. Mainly arboreal, but often occurs on boulders when these are present.

Saltuarius wyberba. Girraween NP.

Granite Belt Leaf-tailed Gecko

Saltuarius wyberba

SVL 109mm

Short conical tubercles on flanks; smooth throat; attenuated tip of original tail more than one-third total tail length, very fine with small tubercles; wide, open V-shaped mark between eyes; moderate size.

NET in vicinity of Stanthorpe and Girraween NP, e. to Queen Mary Falls in adj. sthn BB. Also adj. NSW uplands. Rock-inhabiting, sheltering by day in narrow vertical clefts, emerging at night to rest on boulder faces. Often found exposed at seemingly intolerably low temperatures.

Thick-tailed Geckos

Genus *UNDERWOODISAURUS* 2 Aust. spp.; 1 occurs in Qld.

Thick-tailed or Barking Gecko

Underwoodisaurus milii

SVL 96mm

Original tail plump, carrot-shaped with tapering tip and pale bands; when regenerated, thick, blunt and mottled; plump body with small, granular scales mixed with low tubercles; an area of minute scales on side of snout behind nostril; large head; slender limbs; short, clawed digits with enlarged, transverse subdigital lamellae but no expanded pads. Pink to purplish brown, vividly patterned with small cream spots, each centred on an enlarged tubercle.

SEQ and sthn BB. Also sthn Aust., in well-drained areas of all mainland states. Terrestrial, sheltering by day in burrows or under rocks and logs. Threat response similar to closely related knob-tailed geckos (*Nephrurus*), raising body, and lunging while uttering a loud rasping squeak.

Underwoodisaurus milii. Carnarvon NP.

Genus *UVIDICOLUS* Monotypic genus.

Granite Belt Thick-tailed Gecko

Uvidicolus sphyrurus

SVL 70mm

Original tail plump, squarish with tapering tip and about 4 pale bands; when regenerated, thick, blunt and simply mottled; body plump with small, granular scales mixed with low tubercles; no minute scales on side of snout behind nostril; large head; slender limbs; short, clawed digits with enlarged, transverse subdigital lamellae but no expanded pads. Brownish grey with darker mottling and small pale spots, each centred on a tubercle. Threat response similar to knob-tailed and thick-tailed geckos.

NET, in the vicinity of Girraween and Sundown NPs and Stanthorpe. Also adj. uplands of NSW. Confined to a region subject to cold winters and occasional snow. Terrestrial, sheltering by day in burrows or under rocks and logs.

Uvidicolus sphyrurus. Girraween NP.

AUSTRAL GECKOS

The Northern Phasmid Gecko (*Strophurus taeniatus*) lives only in spinifex hummocks, where its striped pattern aids camouflage and the broadly padded digits ensure a secure grip on the tough slender foliage. Lake Moondarra, Mt Isa area.

Family **Diplodactylidae**

This gecko family is endemic to Australia, New Zealand and New Caledonia. There are 102 described Australian species, with 43 in Queensland. They share important characteristics with the flap-footed lizards (Pygopodidae) and two other Australian gecko families, Gekkonidae and Carphodactylidae.

Immoveable transparent spectacles (brilles) cover the eyes; eye surfaces are cleaned with their broad flat tongues; and they lay just two eggs per clutch. These are soft and parchment-like. Nearly all species have subapical lamellae enlarged to varying degrees and many have digits expanded to form pads. Just one species, the Beaded Gecko (*Lucasium damaeum*), has minute sub-digital scales with no enlarged lamellae.

This extremely diverse family includes terrestrial, arboreal and rock-inhabiting members. Several are restricted to spinifex hummocks, some shelter only in vertical insect and spider holes, while others readily occupy human dwellings.

KEY TO *GENERA*

1 No enlarged scales under digits **(a)** – at most a pair of barely discernible plates under tip.. 2
Two or more enlarged plates under digits, or subdigital scales broad, extending full width of digit................................ 3

2(1) Rostral and mental scales protrude to form a short 'beak'...... .... ***Rhynchoedura***
Snout rounded.................................... ***Lucasium*** (part)

3(1) Claws absent....................................... ***Crenadactylus***
Claws present.. 4

4(3) Enlarged lamellae not terminating in a pair of sub-apical plates; adhesive lamellae under original tail-tip................. .... ***Pseudothecadactylus***
Enlarged lamellae terminating in a pair of sub-apical plates; no adhesive lamellae under original tail-tip..................... 5

5(4) Enlarged lamellae single at base and in paired series under digit **(b)**.. 6
Enlarged lamellae in single series under digit **(c)**, or comprising rounded granules **(d)**................................ 8

6(5) Dorsal scales large, about the same size as ventrals.. ***Oedura***
Dorsal scales minute and normally much smaller than ventral scales... 7

7(6) A series of large, straight-edged, roughly squarish pale dorsal blotches arranged in a ladder-like pattern between broad dark stripes from behind each eye onto to body and original tail............... ***Nebulifera***
A zigzagging pale dorsal stripe, or ragged-edged dorsal blotches with narrow, irregular dark margins........ .... ***Amalosia***

8(5) Digits very short, broad, greatly depressed; lamellae very large and prominent, with very large subapical plates... ***Strophurus***
Digits moderate to narrow, weakly depressed; lamellae moderate to greatly reduced, with moderate to very small subapical plates... 9

9(8) Digits moderate, about 4 times as long as wide; generally more than 5 spines on each side of cloaca......... .... ***Diplodactylus***
Digits narrow, about 7 times as long as wide; clusters of 2–5 spines on each side of cloaca..................... .... ***Lucasium*** (part)

Slender Velvet Geckos
Genus *AMALOSIA*

4 Aust. spp.; 3 occur in Qld. Digits flattened and expanded to form pads, with enlarged subdigital lamellae single at base, paired along digit and ending with a pair of large plates; dorsal scales minute and granular, much smaller than ventrals; weakly dorsally-depressed heads and bodies; tails long and round to weakly depressed in cross-section; irregular pale dorsal blotches or zigzagging vertebral stripe.

Most of Qld, in dry forests, woodlands and rock outcrops. Arboreal and rock-inhabiting, with their flattened builds allowing access to narrow crevices behind bark or under slabs.

KEY TO *AMALOSIA*

1	A zigzagging vertebral stripe; original tail cylindrical	***rhombifer***
	Dorsal blotches (occasionally confluent to form an irregular stripe); original tail depressed	2
2	Little basal webbing between 3rd and 4th toes	***lesueurii***
	Pronounced basal webbing between 3rd and 4th toes	***jacovae***

Amalosia jacovae. Redland Bay.

Amalosia lesueurii. Girraween NP.

Clouded Gecko
Amalosia jacovae
SVL 61mm

Wavy, dark-edged pale grey dorsal zone; pronounced webbing between 3rd and 4th toes; long, slightly bulbous, moderately flattened tail. Grey with paler dorsal zone usually broken or partly broken into blotches by narrow, irregular dark bands.

Endemic. SEQ, in dry eucalypt forests from Brisbane area n. to Kroombit Tops and w. to Warwick area.

Lesueur's Velvet Gecko
Amalosia lesueurii
SVL 80mm

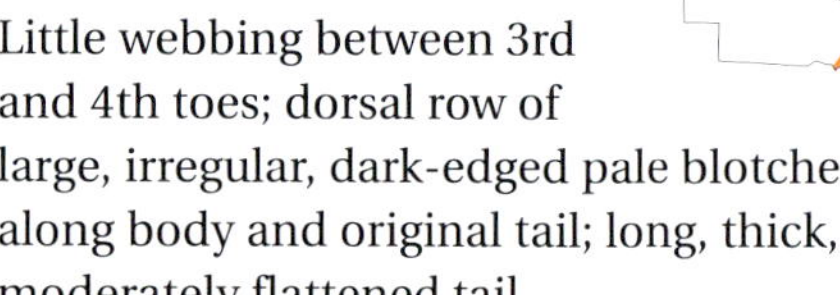

Little webbing between 3rd and 4th toes; dorsal row of large, irregular, dark-edged pale blotches along body and original tail; long, thick, moderately flattened tail.

Rock-inhabiting. NET. Also estn NSW. During mild weather, often shelters under thin slabs strewn over exposed rock surfaces. During summer the more insulated crevices associated with larger boulders are favoured.

Amalosia rhombifer. Eidsvold.

Zigzag Velvet Gecko
Amalosia rhombifer
SVL 70mm

Pale zigzagging vertebral stripe along body and original tail; slender body; long tail, nearly circular in cross-section; Shades of grey or brown with cream to pale grey stripe.

Widespread in variety of forest types over nthn and estn Qld, s. to nthn SEQ. Also NT, nthn WA. Usually arboreal; sometimes associated with rock outcrops and buildings.

Clawless Geckos
Genus *CRENADACTYLUS*

7 Aust. spp.; 2 probably occur in Qld. No claws; very broad digital pads; enlarged but irregular subdigital lamellae, with two very large terminal plates; very small size. Grey to brown with simple darker stripes on back and flanks.

Associated with spinifex, rarely venturing far from cover. Very poorly known in Qld with 2 pops. located at estn extremes of 2 known NT distributions, and tentatively assigned to those spp.

Crenadactylus horni. Alice Springs, NT.

G. Schmida

Crenadactylus naso. Riversleigh.

Central Uplands Clawless Gecko
Crenadactylus horni
SVL 35mm

Internasal scale enlarged; rostral scale fully contacts nostril. Tan to dark brown with simple stripes; usually a narrow pale vertebral, darker paravertebrals, pale dorsolaterals and dark upper laterals and midlaterals (based on NT specimens).

Recorded from thick leaf litter and dense riparian vegtn in Ethabuka area, CHC. Also sthn NT where normally associated with spinifex. Tentatively assigned to *C. horni* based on fragmented Central Aust. occurrence.

Northern Clawless Gecko
Crenadactylus naso
SVL 31mm

Internasal scale not enlarged; rostral scale narrowly contacts nostril; enlarged tubercles on original tail. Brown with simple stripes; pale vertebral, darker paravertebrals, and dark-edged pale dorsolaterals.

Recorded from spinifex on rock at Riversleigh, NWH. Also nthn NT and Kimberley region, WA. Qld pop. tentatively assigned to *C. naso.*

Genus *DIPLODACTYLUS*

27 Aust. spp.; 7 occur in Qld. Highly variable; digits padded, about 4 times as long as wide, with subdigital lamellae ranging from a broad row terminating in two large plates to very small with a minute pair of subapical plates; clusters of usually more than 5 spines on each side of cloaca; tails range from moderately slender and round in cross-section with small uniform scales, to short, plump and flat, encircled by whorls of large squarish scales.

Terrestrial inhabitants of dry, usually open terrain. They shun moist habitats such as RF and are most successful in arid inland areas. One species (*D. vittatus*) thrives in dry sclerophyll forests and throughout SEQ. By day they hide under small surface objects, in burrows and soil cracks.

KEY TO *DIPLODACTYLUS*

a **b**

1	Labial scales large and distinct (**a**); tail cylindrical to moderately depressed	2
	All or most labial scales granular and undifferentiated from adjacent scales (**b**); tail short and greatly depressed	3
2(1)	Back pattern dominated by prominent pale vertebral stripe or equivalent blotches; belly unmarked	***vittatus***
	Back pattern comprises diffuse dark blotches and pale spots; belly with dark blotches	***tessellatus***
3(1)	First upper labial scale small, undifferentiated from remaining upper labial scales; no prominent pale streak from eye to snout	4
	First upper labial scale greatly enlarged, much larger than remaining small granular upper labial scales; prominent pale stripe from eye to snout	5
4(3)	Snout bluntly rounded, broad and U-shaped from above	***ameyi***
	Snout sharply pointed, narrow and V-shaped from above	***platyurus***
5(3)	Mid-dorsal scales small, only slightly larger than adjacent dorsolateral scales	***barraganae***
	Mid-dorsal scales significantly larger than adjacent dorsolateral scales	6
6(5)	Scales on nape plate-like, significantly larger than granular scales on sides of neck; an attenuated extension on original tail-tip	***laevis***
	Scales on nape granular, not significantly larger than scales on sides of neck; original tail blunt with no attenuate tip	***conspicillatus***

Eastern Deserts Fat-tailed Gecko

Diplodactylus ameyi

SVL 60mm

Short, thick spade-like original tail with blunt point but no attenuate tip; scales on dorsal surface of tail in transverse rows, usually uniform but sometimes incl. alternating rows of large and small scales; mid-dorsal

Diplodactylus ameyi. Currawinya NP

body scales body plate-like, noticeably larger than adj. dorsolateral scales; scales on nape granular, just slightly larger than granules on side of neck; all upper labial scales small and granular (1st not enlarged); snout bluntly rounded, broad and U-shaped from above. Shades of brown with fine dark reticulum or extensive dark pigment, generally reduced on vertebral zone, with variable pale spotting. Little or no pale streak from eye to snout.

Widespread in dry to arid woodlands and shrublands on various hard to sandy substrates in ML, CH and MD. Also wstn NSW. Shelters in vertical shafts of insect and spider holes.

J. De Jong

Diplodactylus barraganae. Doomagee.

Gulf Fat-tailed Gecko

Diplodactylus barraganae

SVL 49mm

Short, thick spade-like original tail with blunt tip; scales on dorsal surface of tail in transverse rows, incl. alternating rows of large and small scales; mid-dorsal scales on body small and only slightly larger than adj. dorsolateral scales; 1st upper labial scale greatly enlarged with remainder small and granular; prominent broad pale stripe from eye to snout. Shades of brown with continuous to broken pale vertebral zone, pale spotting and weak dark reticulations.

Recorded from open woodland on red soil. NWH and GUP. Also ne. NT, in broad band along sthn edge of Gulf of Carpentaria. Poorly known. Probably shelters in vertical shafts of insect and spider holes.

Diplodactylus conspicillatus. Emily Gap NT.

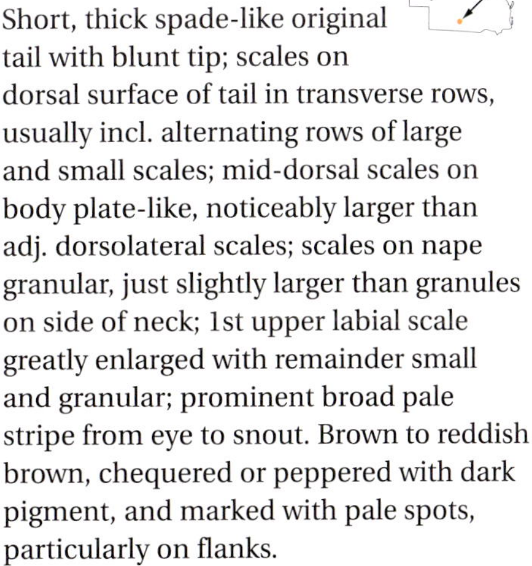

Variable Fat-tailed Gecko

Diplodactylus conspicillatus

SVL 62mm

Short, thick spade-like original tail with blunt tip; scales on dorsal surface of tail in transverse rows, usually incl. alternating rows of large and small scales; mid-dorsal scales on body plate-like, noticeably larger than adj. dorsolateral scales; scales on nape granular, just slightly larger than granules on side of neck; 1st upper labial scale greatly enlarged with remainder small and granular; prominent broad pale stripe from eye to snout. Brown to reddish brown, chequered or peppered with dark pigment, and marked with pale spots, particularly on flanks.

Dry open habitats, mainly on clay, stony, and other hard substrates in NWH, ML and CHC. Also arid areas across SA, NT and WA. Shelters in vertical shafts of insect and spider holes.

Diplodactylus laevis. Nappa Merrie Stn.

Diplodactylus platyurus. Capella.

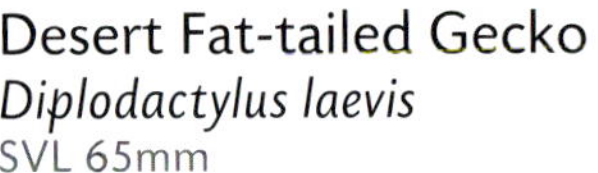

Desert Fat-tailed Gecko

Diplodactylus laevis

SVL 65mm

Thick spade-like original tail with acute attenuate tip; scales on dorsal surface of tail in transverse rows, incl. alternating rows of large and small scales; mid-dorsal scales on body plate-like, noticeably larger than adj. dorsolateral scales; scales on nape and top of head plate-like, considerably larger than granules on side of neck; 1st upper labial scale greatly enlarged with remainder small and granular; prominent broad pale stripe from eye to snout. Shades of brown with dark reticulum and usually fine pale spots, particularly on flanks.

Open shrublands on hard, clay soils in CHC. Also arid zones across SA, NT and WA. Shelters in vertical shafts of insect and spider holes.

Eastern Fat-tailed Gecko

Diplodactylus platyurus

SVL 55mm

Short, thick spade-like original tail with blunt point but no attenuate tip; scales on dorsal surface of tail in transverse rows, usually uniform but sometimes incl. alternating rows of large and small scales; mid-dorsal body scales plate-like, noticeably larger than adj. dorsolateral scales; scales on nape granular, just slightly larger than granules on side of neck; all upper labial scales small and granular (1st not enlarged); snout sharply pointed, narrow and V-shaped from above. Shades of brown with fine dark reticulum or extensive dark pigment, generally reduced on vertebral zone, with variable pale spotting. Little or no pale streak from eye to snout.

Widespread in dry to arid woodlands and shrublands on various hard to sandy substrates in CYP, GUP, EIU, BB, DEU and MGD. Shelters in vertical shafts of insect and spider holes.

Tessellated Gecko

Diplodactylus tessellatus

SVL 50mm

Short tail with rings of large conical scales; large granular body scales; slightly enlarged subdigital lamellae terminating with 2 large plates; rostral scale contacting nostril. Grey to reddish brown with dark mottling, and pale spots typically in 2 paravertebral rows. Belly usually dark-blotched.

Diplodactylus tessellatus. Glenmorgan.

Diplodactylus vittatus. Vertebral stripe straight-edged. Glenmorgan.

Mainly clay soils in BB, ML, CHC, DEU, MGD, possibly sthn GUP. Dry parts of NSW, Vic, SA. Shelters in soil cracks and vertical shafts of disused insect and spider burrows.

Eastern Stone Gecko
Diplodactylus vittatus
SVL 50mm

Fine granular body scales; 2 rows of large subdigital lamellae terminating with 2 relatively large plates; rostral scale contacting nostril. Brown or grey with prominent, straight-edged to deeply notched vertebral stripe or row of blotches, pale spots on flanks.

SEQ, NET, BB, sthn MGD, inland to about Idalia. Also NSW, Vic, SA. Extends well into forested areas more readily than other *Diplodactylus*. Shelters under small surface objects, in burrows, soil cracks. At night often selects slightly elevated perches such as fallen twigs.

Diplodactylus vittatus. Vertebral stripe deeply notched to broken. Sundown NP.

Genus *LUCASIUM*

12 Aust. spp. 6 occur in Qld. Digits narrow, about 7 times as long as wide, with subdigital lamellae normally reduced, ranging from minute and spinose with no subapical plates to moderately enlarged with a pair of subapical plates; clusters of 2–5 spines on each side of cloaca; tails relatively long and slender.

Swift, terrestrial inhabitants of dry, usually open terrain, from woodlands to deserts. By day most species hide in vertical shafts of spider and insect burrows, occasionally soil cracks and less frequently surface objects.

KEY TO *LUCASIUM*

a b

1 All subdigital scales minute; no indication of enlarged plates **(a)** .... ***damaeum***
A pair of plates (either large and conspicuous or small and just discernible) beneath tips of digits **(b)**; 2

2(1) Body scales small mixed with larger rounded tubercles .... ***byrnei***
Body scales small and uniform 3

3(2) Rostral scale contacts nostril. 4
Rostral scale separated from nostril 5

4(3) Digits expanded to form pads; a pair of obvious plates beneath tips of digits. .... ***iris***
Digits narrow, not expanded; plates beneath tips of digits small and just discernible .... ***microplax***

5(3) Pale vertebral stripe encloses darker 'islands', or broken to form blotches. .... ***steindachneri***
Pale vertebral stripe solid .... ***immaculatum***

Gibber Gecko
Lucasium byrnei
SVL 55mm

Fine body scales mixed with larger rounded tubercles; long slender tail; granular subdigital lamellae with moderately large terminal plates; rostral scale separated from nostril. Yellowish brown to reddish brown with about 4 large, irregular, paler dorsal blotches.

CHC. Open shrublands on heavy to stony soils. Also inland NSW, SA, sthn NT. Recorded as sheltering and laying eggs in vertical shafts of spider holes.

Lucasium byrnei. Betoota district.

Beaded Gecko
Lucasium damaeum
SVL 55mm

Fine granular body scales; minute spinose subdigital lamellae with no enlarged terminal plates; rostral scale contacting nostril. Reddish brown with broad, pale, deeply notched to straight-edged vertebral stripe or row of blotches. Flanks prominently marked with pale spots.

Dune slopes and sandy flats, usually with spinifex, in CHC. Also arid zones across sthn Aust. Shelters in insect and spider holes, foraging at night on exposed sandy areas. Alert; quick to flee in torchlight.

Lucasium damaeum. Hattah-Kulkyne NP, Vic.

Lucasium immaculatum. Mt Isa, Qld.

Pale-striped Ground Gecko

Lucasium immaculatum

SVL 85mm

Fine granular body scales; moderately enlarged granular subdigital lamellae with 2 large terminal plates; rostral scale separated from nostril. Reddish brown with narrow pale stripes extending from snout through eyes to join on nape, forming prominent vertebral stripe continuous with series of narrow pale vertical bars on flanks.

Heavy to stony soils with open shrublands in CHC, NWH, MGD. Also estn NT.

E. Vanderduys

Lucasium iris. Gilberton Stn.

Gilbert Ground Gecko

Lucasium iris

SVL 62mm

Fine granular body scales; digits expanded to form pads with enlarged subdigital lamellae and 2 large terminal plates; rostral scale contacting nostril. Brown to reddish brown with broad, prominent pale vertebral stripe with straight or zigzagging edges expanding over top of head and extending back to base of tail, scattered small pale spots on back and flanks, and irregular pale blotches along top of original tail.

Endemic. Low sandstone hills with Acacia and eucalypt woodlands over spinifex in Gregory Ra. on Gilberton Stn and Rungulla NP in EIU.

Lucasium microplax. Windorah.

Southern Sandplain Gecko

Lucasium microplax

SVL 55mm

Fine granular body scales; digits not expanded to form pads, with slightly enlarged subdigital lamellae and 2 very small, barely discernible terminal plates; rostral scale contacting nostril. Dark reddish brown with prominent pale vertebral stripe forking at nape and extending onto base of original tail. Flanks have moderately large pale spots, some separated and others confluent with each other or vertebral stripe.

Open, mainly sandy areas, often with spinifex, but also clay or stony sites in CHC. Also sthn NT, SA and estn WA.

Lucasium steindachneri. Blotched pattern. Coen.

Lucasium steindachneri. Unbroken dorsal pattern. Currawinya NP.

Box-patterned Gecko

Lucasium steindachneri

SVL 55mm

Fine granular body scales; slightly enlarged subdigital lamellae terminating with 2 large plates; rostral scale separated from nostril. Shades of brown with broad pale vertebral zone broken into angular blotches or partly broken to enclose up to 4 dark oval 'island' patches on midline.

Variety of vegtn and soil types in CHC, BB, ML, MGD, DEU, GP, EIU, CYP, n. to Coen area. Also NSW, far estn SA. Occasionally hides under timber or small stones but normally shelters deeper in insect holes and soil cracks.

Genus *NEBULIFERA* Monotypic genus.

Robust Velvet Gecko

Nebulifera robusta

SVL 80mm

Digits flattened and expanded to form pads, with enlarged subdigital lamellae single at base, paired along digit and ending with a pair of large plates; dorsal scales minute and granular, much smaller than ventrals; head and body weakly dorsally-depressed; tail thick and flattened, sometimes extremely broad when regenerated. Shades of brown to blackish brown with large, squarish, pale dorsal blotches along back and original tail.

Dry forests, woodlands and rock outcrops in SEQ, NET and sthn BB. Also nthn NSW. Arboreal and rock-inhabiting; commonly occupies buildings in rural areas. Though usually absent in inner urban areas, large pop. thrives on the Kangaroo Point cliffs in central Brisbane.

Nebulifera robusta. Kurwongbah.

Velvet Geckos
Genus *OEDURA*

19 Aust. species; 11 occur in Qld. Digits flattened and expanded to form pads, with enlarged subdigital lamellae single at base, paired along digit and ending with a pair of large plates; dorsal scales smooth, uniform, large and flat, no smaller than ventral scales; weakly dorsally-depressed heads and bodies; tails long, usually fleshy and depressed; normally boldly patterned, brightest on juv.

Throughout Qld, living on tree trunks and rock faces. Flattened build allows access to narrow crevices behind bark or under slabs. Some commonly occupy buildings. Often found associated with sloughed skins of various ages, suggesting long term occupancy of shelter sites.

KEY TO *OEDURA*

1	Body banded, spotted, blotched or with short bars	2
	Body mostly without pattern; markings largely restricted to bands across neck and base of tail	***jowalbinna***
2(1)	Body banded; or if fragmented then 2 or more cloacal spurs	3
	Body spotted, blotched, striped or with short bars; one cloacal spur	6
3(2)	Pale bands form deep U-shapes	***castelnaui***
	Pale bands either transverse, forming shallow U-shapes or fragmented	4
4(3)	Rostral scale fully divided	***cincta***
	Rostral scale partly divided	5
5(4)	One cloacal spur	***argentea***
	2–3 cloacal spurs	***bella***
6(2)	Pattern blotched or striped; if bars then no white spots on limbs	7
	Pattern spotted or with short bars; limbs usually spotted, at least hindlimbs	10
7(6)	Blotches or stripes in roughly paired series along either side of midline	8
	Blotches dispersed across dorsal surface	***picta***
8(7)	V- or Y-shaped broken bar on nape; spots, blotches or lines arranged linearly alongside thin, pale midline	***lineata***
	Complete curved pale bar across nape; dorsal blotches in paired series, often joined to form dumbbell shapes	9
9(8)	Thin black line from below eye to below nape band; black edging to pairs of dorsal markings and nape bar does not extend to other pairs of markings	***elegans***
	Dark band from back of eye to nape band; black edging to pairs of dorsal markings and nape bar connecting at least anterior pairs of markings	***monilis***
10(6)	Spots tend to join forming transverse bars, usually including curved band on nape; usually dark band from eye to nape or broken on temporal region; nthn Qld	***coggeri***
	Spots rarely joined; nape spotted or blotched but no continuous bar; sthn to mid-eastern Qld	***tryoni***

Silver-eyed Velvet Gecko

Oedura argentea

SVL 80mm

Transverse pale bands with straight dark edges; moderately short narrow tail; partially divided rostral scale. Purplish grey and yellow with 5–6 pale bands across body, not narrowing on flanks. Nuchal band sweeps forward unbroken through ear and lips to snout.

Endemic. Rock-inhabiting, on sandstones in Gregory Range and Bulleringa NP areas EIU.

Oedura argentea. Cobbold Gorge.

Gulf Marbled Velvet Gecko

Oedura bella

SVL 92mm

Variably banded to spotted; moderately short tail; partially divided rostral scale. Dark purplish brown with about 5 cream to bright yellow bands across body. Bands prominent, sharp edged and simple on juv, fragmenting with age to become ragged-edged, and dark pigment within pale bands, and pale mottling within dark interspaces. Bands may be completely replaced by an irregular mosaic of pale spots and blotches.

Rock-inhabiting in NWH. Also ne. NT.

Oedura bella. Lawn Hill NP.

Northern Velvet Gecko

Oedura castelnaui

SVL 90mm

Prominent pale bands almost as wide as or wider than dark interspaces; large size; thick, slightly flattened tail. Pale bands (about 5 on body and 5 on original tail), cream alternating with yellow and dark brown to black.

Endemic. Dry forests, woodlands, rock outcrops of nthn BB, EIU, WT, CYP, n. to sthn Torres Strait is. Arboreal, normally sheltering under loose bark; also recorded under dry skirts of grass trees (*Xanthorrhoea spp.*).

Oedura castelnaui. White Mtn NP.

Oedura cincta. South Galway Stn.

Oedura coggeri. Undara NP.

Inland Marbled Velvet Gecko

Oedura cincta

SVL 108mm

Variably banded to spotted; moderately long tail; rostral scale fully divided by rostral crease; tail roughly circular in cross-section. Dark purplish brown with dark band from behind eye, across nape and about 5–6 pale grey bands across body. Bands prominent, sharp edged and simple on juv, but fragmenting with age, becoming ragged-edged with dark centres, and pale spots form within dark interspaces. Bands may be completely overridden by an irregular mosaic of pale spots and blotches.

Widespread through BB, ML, CHC, MGD and DEU. Also dry areas of NSW, SA and sthn NT. Dry open forests and rock outcrops. Mainly arboreal but also exploiting rocky habitats.

Northern Spotted Velvet Gecko

Oedura coggeri

SVL 70mm

Spotted; prominent dark-edged pale spots tending to fuse into short transverse bars, usually beginning with pale band across nape; short, slightly flattened tail. Brownish yellow with dark-edged cream to pale grey spots and/or bars. Limbs spotted.

Endemic. WT, EIU. Rock outcrops assoc. with dry open woodlands. Largely rock-inhabiting, less frequently dwelling on trees.

Elegant Velvet Gecko

Oedura elegans

SVL 89mm

Dorsal row of circular pale blotches, arranged as pairs or narrowly joined to form dumbbell shapes on midline, along body and original tail; moderate, rounded tail; partially divided rostral scale. Yellowish brown with pale U-shape on nape, a thin dark line extending from below eye to below nape marking, and paired cream to pale grey dorsal blotches, each dark-edged but dark pigment not extending to join other pairs.

Dry timbered areas, particularly where ironbarks and cypress pines are prominent in sthn BB, Also nthn NSW. Usually arboreal; occasionally rock-inhabiting.

Oedura elegans. Amby area, Qld.

Quinkan Gecko

Oedura jowalbinna

SVL 69mm

Back mostly patternless; pattern largely restricted to nape and hips. Pinkish grey with dark-edged pale yellow band across neck continuous with pale stripe through lip and under eye, and short pale band across hips.

Endemic. Jowalbinna Station, CYP, on deeply dissected escarpments of Laura Sandstone. Inhabits cliffs and overhangs.

E. Vanderduys

Oedura jowalbinna. Jowalbinna Stn.

Arcadia Velvet Gecko

Oedura lineata

SVL 79mm

Blotched; pale midline; relatively long rounded tail; partially divided rostral scale. Yellow variegated with black, with pale V- or Y-shaped nuchal band, broken on midline, and a series of dark-edged pale dorsal blotches or spots, sometimes extended longitudinally to form short stripes, along either side of thin yellowish vertebral line.

Endemic. Restricted to brigalow forest in Arcadia Valley between Expedition NP and Nuga Nuga NP in BB. Arboreal, sheltering under exfoliating brigalow bark. Limited pops now fragmented due to clearing.

Oedura lineata. Arcadia Valley.

Ocellated Velvet Gecko

Oedura monilis

SVL 96mm

Dorsal row of circular pale blotches, arranged as pairs or narrowly joined to form dumbell shapes on midline; moderate, rounded tail; partially divided rostral scale. Yellowish brown with pale U-shape on nape, broad dark band from rear of eye to nuchal band, and paired cream to pale grey dorsal blotches, each dark-edged, with dark pigment usually extending to join other pairs at least anteriorly.

Endemic. Woodlands to dry eucalypt forests, dry vine thickets and rock outcrops in nthn BB and EIU. Arboreal and rock-inhabiting.

Oedura monilis. Mt Cooper Stn.

Ornate Velvet Gecko

Oedura picta

SVL 79mm

Blotches spread irregularly across back and flanks; relatively long rounded tail; partially divided rostral scale. Yellow with dark flecking and mottling, a dark edged pale bar across nape and large, dark-edged pale spots and blotches spread irregularly across back, becoming more fragmented on flanks, and dark band from snout, below eye to top of ear but not reaching nuchal bar.

Endemic. Rock-inhabiting, on sandstone outcropping in Denham Range, Dysart region, BB.

Oedura picta. Bundoora State Forest.

Oedura tryoni. Spring Gully area.

Southern Spotted Velvet Gecko

Oedura tryoni

SVL 87mm

Numerous small, dark-edged pale spots over all dorsal surfaces; thick, slightly flattened tail. Shades of brown to yellowish brown with cream spots margined with black. Limbs spotted. Nape spotted or blotched but no continuous bar. Nthn pop. in CQC has larger spots forming blotches.

SEQ, NET, BB and CQC. Also nthn NSW. Arboreal and rock-inhabiting. Occupies variety of habitats, from exposed exfoliating granite slabs to inland stands of brigalow and coastal woodlands.

Oedura tryoni. Eungella NP

Genus *PSEUDOTHECADACTYLUS*

3 Aust. spp.; 1 occurs in Qld.

Giant Tree Gecko

Pseudothecadactylus australis

SVL 120mm

Cylindrical prehensile tail with ridges of adhesive lamellae beneath tip; uniform granular scales; robust body with deep head; digits very broad and flattened to form pads, with wide paired rows of lamellae under entire length; all digits clawed except innermost. Brown to grey with large pale dorsal blotches, often divided into pairs, particularly on nape.

Endemic. Nthn CYP to sthn Torres Strait is. Arboreal, dwelling in woodlands, monsoon forests, mangroves. Shelters by day in hollows, forages at night on branches and vines. Agile climber, employing prehensile tail like 5th limb.

Pseudothecadactylus australis. Lockerbie Scrub.

Beaked Geckos Genus *RHYNCHOEDURA*

6 Aust. spp.; 4 occur in Qld. Short beak-like snout; small granular labial scales; long slender body; short clawed digits; subdigital lamellae small and granular, ending with a very small pair of plates. Most spp. extremely similar; only identified with certainty using genetics. Locality also important.

Widespread throughout dry interior. Also arid zones of all mainland states. Occur mainly on red soils in most open habitats, particularly shrublands and spinifex grasslands. They mainly shelter in the vertical shafts of spider holes, and feed almost exclusively on termites.

KEY TO *RHYNCHOEDURA*

1 Three enlarged scales under chin (**a**); 2 preanal pores separated by smaller scales (**b**) .... ***mentalis***
Five enlarged scales under chin (**c**); 2 preanal pores in contact (**d**) 2

2(1) Pale dorsal spots tend to form elongate blotches; far sw. Qld. .... ***eyrensis***
Pale dorsal spots generally separate; Qld interior 3

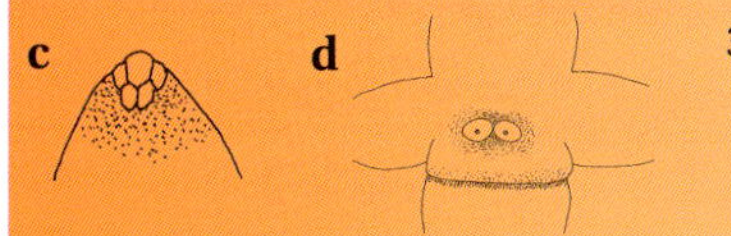

3(2) Pale dorsal spots medium-sized; mental scale rounded; most of Qld interior .... ***ormsbyi***
Pale dorsal spots fewer and smaller; mental scale elongate; narrow band of Qld interior, s. from Longreach...... ***angusta***

Rhynchoedura angusta. Noonbah Stn.

Border Beaked Gecko
Rhynchoedura angusta
SVL 51mm

Five enlarged scales under chin; 2 enlarged preanal pores in contact; single cloacal spur; rostral scale with median groove. Reddish brown with pale, wavy-edged vertebral zone tending to form transverse bars. Head and body with numerous, small to medium, similarly-sized white to pale yellow spots. MGD and ML, in narrow band of alluvial flood plains of Bulloo-Bancannia drainage division n. to Longreach. Also wstn NSW. Mitchell grass and spinifex.

Eyre Basin Beaked Gecko
Rhynchoedura eyrensis
SVL 51mm

Five enlarged scales under chin; 2 enlarged preanal pores in contact; single cloacal spur; rostral scale with median groove. Reddish brown with pale, wavy-edged vertebral zone sometimes forming transverse bars. Head and body with numerous, white to pale yellow spots often merged to form elongated blotches and tending to be largest on flanks.

CC, on sandy and stony terrain. Also arid zones across Lake Eyre Basin in estn SA.

Rhynchoedura eyrensis. Nappa Merrie Stn.

Rhynchoedura mentalis. Cooladdi area.

Rhynchoedura ormsbyi. Morven.

Brigalow Beaked Gecko
Rhynchoedura mentalis
SVL 50mm

Three enlarged scales under chin; two enlarged preanal pores separated by several scales; cloacal spurs comprise 2–3 enlarged scales; no groove in rostral scale. Dark reddish brown with straight-edged tan vertebral stripe, or with dark mottling and transversely elongate pale bars. Distinctive pale iris.

Endemic. MGD and ML, n. to Longreach area and w. to Eromanga area. Mulga and shrublands on clay to stony soils.

Eastern Beaked Gecko
Rhynchoedura ormsbyi
SVL 50mm

?

Five enlarged scales under chin; 2 enlarged preanal pores in contact; single cloacal spur; rostral scale with median groove. Reddish brown with pale, wavy-edged vertebral zone and numerous pale yellow spots, not normally forming transverse bars and tending to be largest on flanks.

Widespread over estn dry to arid zones. BB, ML, DEU, EIU, GUP, MGD, and possibly NWH and CHC. Also wstn NSW. Open habitats incl. alluvial flats, shrublands, eucalypt and mulga woodlands.

Spiny-tailed, Striped and Jewelled Geckos
Genus *STROPHURUS*

20 Aust. species; 8 occur in Qld. Body scales either fine and homogenous, or with 2–4 rows of large, rounded tubercles or spines, or with scattered, slightly enlarged scales; digits short and flattened to form pads, with enlarged subdigital lamellae divided at base, single along digit, and ending with two large plates.

Widespread in dry parts of the state, extending east to BB. Also dry parts of all mainland states. They share the unique ability to squirt a sticky, thick repellent fluid from glands along the back and tail when under duress. Included in this variable group are extremely distinctive species with brightly coloured irises. Some are arboreal, others are specialised spinifex inhabitants.

KEY TO *STROPHURUS*

1	Enlarged spines or tubercles present, prominent and protrusive	2
	No enlarged spines; tubercles absent or very low and inconspicuous	4
2(1)	Spines over eye; 2 long rows of spines along top of original tail	3
	No spines over eye; 4 rows of tubercles along top of original tail	***williamsi***
3(2)	Mouth lining yellow; 4–7 granular scale rows between rows of caudal spines	***ciliaris***
	Mouth lining blue; 8–13 granular scale rows between rows of caudal spines	***krisalys***
4(1)	Simple stripes along upper and lower surfaces	***taeniatus***
	Pattern plain, or dominated by spots or a bold black and white reticulum	5
5(4)	Complex black and white mosaic; grey to orange blaze on tail	***taenicauda***
	Pattern obscure or with simple sharp white spots	6
6(5)	Leaden grey with simple sharp white spots	***elderi***
	Pattern obscure	7
7(6)	Iris bright yellow to golden	***trux***
	Iris cream with orange to brown reticulum	***congoo***

Northern Spiny-tailed Gecko

Strophurus ciliaris

SVL 77mm

Long spines above eyes; 2 rows of long spines along top of tail, moderately spaced and separated by 4–7 rows of granular intercaudal scales; large tubercles scattered over back; yellow-orange mouth-lining. Varies from grey with two wavy lines along each side of back enclosing a darker zone, to cream or brown with bright orange blotches. Iris rimmed with orange to maroon, enclosing a vivid dark and pale reticulum.

Widespread in ML, CHC, MGD, NWH, GUP, occupying arid shrublands, sand dunes, tropical woodlands, riverine gorges. Clings by day to stems in sheltered or exposed sites; forages at night amongst foliage or on ground. Two ssp. described across nthn Aust. but status of estn pops (incl. those from Qld) undetermined.

Strophurus ciliaris. Lawn Hill.

Strophurus ciliaris. Nappa Merrie Stn.

E. Vanderduys

Strophurus congoo. Emu Creek.

Congoo Gecko
Strophurus congoo
SVL 49mm

Weakly patterned; no enlarged tubercles or spines; moderately slender; relatively short tail; cream iris with reddish-brown reticulations; pale to dark blue mouth-lining. Cream, tan to dark grey with dark peppering. Flanks and paler ventral surfaces sharply delineated, and often a faint cream upper lateral stripe on tail.

Endemic. EIU, in a small area of infertile granitic and rhyolite hilly country near Petford. Spinifex inhabitant.

Strophurus elderi. Windorah.

Jewelled Gecko
Strophurus elderi
SVL 48mm

Fine granular scales mixed with scattered, slightly enlarged scales; short thick tail. Dark leaden grey with starburst of sharp white spots, each centred on enlarged scale, scattered liberally over head, body, limbs and tail.

CHC, e. to Windorah area. Also sandy deserts in all mainland States except Vic. Lives in spinifex clumps on red sandy ground, seldom straying far from the slender, formidably spiny foliage, in which they climb effortlessly.

Strophurus krisalys. Congie Stn, Eromanga area.

Kristin's Spiny-tailed Gecko
Strophurus krisalys
SVL 70mm

Moderately long spines above eyes; 2 rows of long spines along top of tail, widely spaced and separated by 8–13 rows of granular intercaudal scales; large tubercles scattered over back; blue-grey mouth-lining. Grey with hourglass-shaped lateral blotches in n. and with weaker pattern in s. Iris rimmed with deep maroon, enclosing a vivid dark and pale reticulum.

Endemic (possibly estn SA.). ML, CHC, MGD, NWH and GUP. Shrublands and mulga woodlands on hard stony soils, and spinifex-dominated habitats on sandy soils.

Northern Phasmid Gecko
Strophurus taeniatus
SVL 44mm

Extremely slender body, limbs and tail; fine granular scales with no enlarged tubercles; simple striped

Strophurus taeniatus. Lake Moondarra.

Strophurus taenicauda albiocularis. Dingo area.

pattern. Shades of grey, brown to yellow with straight, sharp-edged stripes along dorsal, lateral and ventral surfaces.

NWH, with isolated estn record at Gilberton Stn, on border of EIU and GUP. Also NT, WA. Lives in spinifex clumps, mainly on heavy red loams and stony soils.

Golden-tailed Gecko

Strophurus taenicauda

SVL 73mm

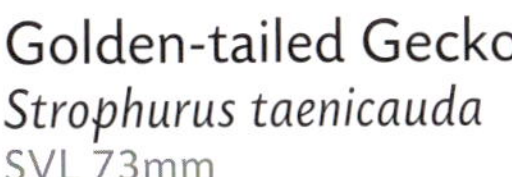

Strikingly coloured with network of black spots against pale grey to cream; orange blaze along top of tail; grey to bright red eye; dark blue mouth-lining; no spines or enlarged tubercles. 3 ssp: *S. t. taenicauda* has bright red iris and 1 caudal stripe with undulating edges or transverse extensions. *S. t. albiocularis* has white to cream iris and 1 straight-edged caudal stripe. *S. t. triaureus* has amber iris and 3 caudal stripes; broad, straight dorsal and thin undulating laterals separated by narrow black and white vertical bars.

Endemic. Sthn BB; in dry sclerophyll forests featuring ironbarks, cypress pine and brigalow. *S. t. taenicauda* occupies s. of range, nw. to Carnarvon NP. *S. t. albiocularis* occurs in n., to sw. of Rockhampton. *S. t. triaureus* occupies small area between Mundubbera, Callide and Biloela. Arboreal, sheltering behind loose dead bark, in hollows, or clinging to exposed slender branches in dappled sunlight.

Strophurus taenicauda triaureus. Eidsvold.

Strophurus taenicauda taenicauda. Meandarra.

Strophurus trux. Marlborough area.

Strophurus williamsi. Bendidee SF.

Golden-eyed Gecko

Strophurus trux

SVL 49mm

Weak pattern; no enlarged tubercles or spines on body; 2–4 slightly enlarged supraciliary tubercles above rear of eye; bright yellow to rich gold or golden brown iris; pale blue mouth-lining with pink to red tongue. Pale grey peppering of small darker spots, occasionally obscure dark dorsal reticulations and often faint dark stripes on tail.

Endemic. Restricted to a small area of steep rocky hills with spinifex near Marlborough in estn BB. Spinifex inhabitant, often climbing into low shrubs at night.

Eastern Spiny-tailed Gecko

Strophurus williamsi

SVL 60mm

Enlarged tubercles present, in 2–4 nearly parallel rows on back, 4 rows on tail; reddish orange rim around iris; bluish purple mouth-lining. Shades of grey with fine dark spots, a fine dark network, and often broad, wavy-edged darker dorsal zone.

Dry sclerophyll forests and woodlands, often featuring ironbark eucalypts, cypress pines, brigalow or mulga. BB, DEU, MGD, ML. Also NSW, far estn SA. Arboreal, sheltering under loose bark, in hollows or in dappled sunlight on exposed slender branches. One seen clinging to cyclone wire fence in Charleville, in direct sunlight at about 40°C.

COSMOPOLITAN GECKOS

The Cooktown Ring-tailed Gecko (*Cyrtodactylus tuberculatus*) is one of Australia's largest species. It has slender clawed digits, with no adhesive pads, yet it is a skilled and agile climber on rocks and vines. Cooktown.

Family **Gekkonidae**

Gekkonid geckos occur world-wide, occupying all continents and many isolated islands. Their distribution encompasses tropical and temperate regions. There are 77 described species in Australia and its external territories, and 24 in Queensland.

Like the flap-footed lizards (Pygopodidae) and the two other Australian gecko families, gekkonids have immoveable transparent spectacles covering the eyes. Eye surfaces are cleaned using their broad flat tongues. Digits range from very broad and flat with rows of large subdigital lamellae bristling with adhesive setae, to slender and bird-like with simple claws.

The unique eggs have brittle calcareous shells, unlike the soft, malleable shells of other lizards' eggs. These are relatively water-proof, so can be laid in dry sites under bark, in cracks or even glued to firm surfaces. Impervious eggs probably play a significant role in the family's cosmopolitan distribution and extraordinary dispersal success. One introduced species, the Asian House Gecko (*Hemidactylus frenatus*), is hailed as the world's most invasive lizard. It has gained a sure foot-hold in Queensland and is dispersing rapidly.

Gekkonids include terrestrial, arboreal and rock-inhabiting species. Many species in Australia and overseas readily occupy human dwellings.

KEY TO GENERA

1	Digits angular, not resting flat on substrate, with obvious terminal claws **(a)**; or at least digits very narrow and not wider than deep	2
	Digit wider than deep and resting flat on substrate	4
2(1)	Two scales beneath base of claw **(b)**	***Heteronotia***
	One broad scale beneath base of claw **(c)**	3
3(2)	A lateral fold of skin between armpit and groin; extremely large	***Cyrtodactylus***
	No lateral fold of skin between armpit and groin; small to moderate	***Nactus***
4(1)	Claws arise conspicuously from well within upper surface of pad **(d)**	5
	Free portion of claw extends from tip of pad **(e)**	***Lepidodactylus***
5(4)	Inner digit clawless; all scales uniformly granular	***Gehyra***
	Inner digit clawed; slightly enlarged tubercles on back, and bands of prominent tubercles on original tail	***Hemidactylus***

Ring-tailed Geckos
Genus *CYRTODACTYLUS*

140+ spp. centred mainly in Asia and Pacific; 6 occur in Aust.; 5 in Qld. Large and banded; clawed, padless somewhat bird-like digits; long slender tails; small granular dorsal scales mixed with longitudinal rows of larger tubercles; a ventrolateral skin fold. Among largest geckos in Aust. Arboreal and rock-inhabiting. Agile climbers on rough surfaces, leaping with ease between boulders.

KEY TO *CYRTODACTYLUS*

1	Forearm with enlarged tubercles	2
	Tubercles on forearm sparse to absent; McIlwraith Range	***pronarus***
2(1)	> 50 enlarged scales from knee to knee along rear ventral edge of thigh anterior to cloaca	***adorus***
	< 50 enlarged scales from knee to knee along rear ventral edge of thigh anterior to cloaca	3
3(2)	Femoral and cloacal pores in 3 separate patches; dark body bands usually with dark posterior edges only	***mcdonaldi***
	Femoral and cloacal pores in continuous series; dark body bands usually with dark posterior and anterior edges	4
4(3)	Pale body bands include dark ventrolateral bars; north-eastern CYP in vicinity of Iron Range	***hoskini***
	Pale body bands without dark ventrolateral bars (occasionally rounded pots present); south-eastern CYP, south of Cape Melville area	***tuberculatus***

Pascoe River Ring-tailed Gecko

Cyrtodactylus adorus

SVL 123mm

Small tubercles on forearm; dorsal tubercles moderately developed in 21–24 longitudinal rows at midbody; 56–65 enlarged scales along rear ventral edge of thigh anterior to cloaca. Prominent, sharp-edged cream and brown bands on body with pale interspaces unmarked, incl. usually 3 dark bands on trunk, each with narrow darker rear edge and often narrow paler anterior edge. Top of head largely patternless. Dark bands on tail-base about twice as wide as pale interspaces.

Endemic. Nthn CYP, near lower reaches of Pascoe River, incl. Wattle, Stanley and Kennedy Hills. Rocky outcrops, mainly granite.

K. Aland

Cyrtodactylus adorus. Kennedy Hills.

Iron Range Ring-tailed Gecko

Cyrtodactylus hoskini

SVL 112mm

Large, moderately projecting tubercles on forearm; dorsal tubercles strongly developed in 19–24 longitudinal rows at midbody; 41–48 enlarged scales along rear ventral edge of thigh anterior to cloaca. Prominent, dark-edged cream and brown bands on body, each with narrow posterior and anterior vertebral extensions. Pale interspaces usually enclose dark lateral patch or bar. Coarse dark mottling on top of head. Dark bands on tail-base slightly wider than pale interspaces.

Endemic. Nthn CYP on wstn edge of Iron Range. Granite boulders in open forest, RF patches, ferns, vines and umbrella trees.

E. Vanderduys

Cyrtodactylus hoskini. Tozer Range.

Southern Ring-tailed Gecko

Cyrtodactylus mcdonaldi

SVL 105mm

Large, weakly to moderately projecting tubercles on forearm; dorsal tubercles strongly developed in 18–23 longitudinal rows at midbody; 34–46 enlarged scales along rear ventral edge of thigh anterior to cloaca.

Cyrtodactylus mcdonaldi. Chillagoe.

Prominent cream and brown bands with dark rear edges on body, no markings in pale interspaces. Top of head smudged with grey. Dark bands on tail-base just slightly wider than pale interspaces.

Endemic. EIU, from Chillagoe n. to Parrot Ck in Shipton's Flat area. Rocky areas incl. limestone at Chillagoe, sandstone at Mt Mulligan and granite at Mareeba.

McIlwraith Ring-tailed Gecko

Cyrtodactylus pronarus

SVL 132mm

No tubercles on forearm; dorsal tubercles moderately developed in 20–24 longitudinal rows at midbody; 58–66 enlarged scales along rear ventral edge of thigh anterior to cloaca. Prominent cream and brown bands on body have dark posterior edges, but diffuse anterior edges merge into paler interspaces. Pale interspaces usually unmarked. Top of head largely patternless. Dark bands on tail-base more than twice as wide as pale interspaces.

K. Aland

Cyrtodactylus pronarus. Peach Creek, McIlwraith Range.

Endemic. Nthn CYP, in small area of McIlwraith Ra at about 500 m altitude. Granite boulders and RF at Peach Ck.

Cyrtodactylus tuberculatus. Black Mountain via Cooktown.

Cooktown Ring-tailed Gecko

Cyrtodactylus tuberculatus

SVL 120mm

Strongly projecting tubercles on forearm; dorsal tubercles strongly developed in 20–24 longitudinal rows at midbody; 34–46 enlarged scales along rear ventral edge of thigh anterior to cloaca. Prominent, dark-edged cream and brown bands on body, often with a few dark spots within pale interspaces. Top of head has strong dark mottling. Dark bands on tail-base just slightly wider than pale interspaces.

Endemic. Sthn CYP from just n. of Cape Melville (Stanley and Flinders Is.), s. to Mt Leswell. Granite boulders and sandstone outcrops, extending onto adjacent vegtn. Occasionally inhabits buildings.

Dtellas

Genus *GEHYRA*

51 spp. in Aust. and external territories; 11 occur in Qld. Other spp. extend from South-East Asia, through NG to wstn Pacific. Flat digits, greatly expanded to form large circular pads; all digits clawed except innermost, the claws arising from the centre of each pad and extending forward well beyond it; greatly expanded subdigital lamellae, forming entire or divided series under circular part of digit; fine uniform body scales; weakly dorsally depressed heads and bodies.

Swift, alert geckos that shelter behind loose bark, in rock crevices and in vertical spaces (behind paintings, etc.) in dwellings. Expert climbers, scuttling with ease over smooth rock faces, tree trunks and walls. Often extremely abundant. Their round, brittle-shelled eggs are frequently laid communally in rock crevices, hollow limbs and fence posts.

KEY TO *GEHYRA*

a b

1	Subdigital lamellae undivided (but may be deeply notched and initially appear divided) **(a)**	2
	Subdigital lamellae divided **(b)**	6
2(1)	Nine or more lamellae under expanded pad on 4th toe	3
	Fewer than 9 lamellae under expanded pad on 4th toe	5
3(2)	Preanal pores 22 or more; little or no pattern; NWH	***lauta***
	Preanal pores 20 or fewer; some indication of pattern including a dark stripe behind eye and/or dark mottling, or dark bands and rows of pale spots	4
4(3)	Greyish colour with dark spotting and mottling, and usually a dark stripe behind eye; widespread	***dubia***
	Orange brown with dark bands alternating with rows of pale spots; NWH	***robusta***
5(2)	Pinkish orange with pale spots and dark blotches	***electrum***
	Brown to grey with dark chain-like dorsal pattern	***catenata***
6(1)	Web of skin on rear of hindlimb ; Torres Strait	***baliola***
	No web of skin on rear of hindlimb	7
7(6)	More than 9 lamellae under expanded pad on 4th toe	***borroloola***
	Nine or fewer lamellae under expanded pad on 4th toe	8
8(7)	Pattern dominated by spots; rock-inhabitant	9
	Spots present or absent but pattern usually dominated by streaks or variegations; inhabits trees, rock and human dwellings	10
9(8)	Subdigital lamellae 6 or fewer; EIU	***einasleighensis***
	Subdigital lamellae 7–8; far wstn CHC	***moritzi***
10(8)	Upper edges of rostral scale weakly to strongly sloping; 11 or more preanal pores in males	***versicolor***
	Upper edges of rostral scale flat or very weakly sloping; usually fewer than 11 preanal pores in males	***purpurascens***

H. Cogger

Gehyra baliola. Murray Is.

Gehyra catenata. Dingo area.

Short-tailed Dtella

Gehyra baliola

SVL 100mm

Subdigital lamellae mostly divided, 12–16 under expanded pad of 4th toe; distinctive web of skin at rear of hindlimb; U-shaped rostral scale enclosing up to 12 small scales; large size. Yellowish brown to grey with paler spots. Ventral surfaces often yellow.

Darnley and Murray Is, ne. Torres Strait. Widespread in sthn NG.

Chain-backed Dtella

Gehyra catenata

SVL 55mm

Subdigital lamellae undivided; 7–8 under expanded pad of 4th toe; no web of skin at rear of hindlimb. Grey with chain-like dorsal row of dark-edged pale blotches.

Endemic. Woodlands (mainly acacia) in BB, DEU, possibly CHC. Arboreal, sheltering under loose bark of dead trees.

Gehyra borroloola. Limmen Bight R., NT.

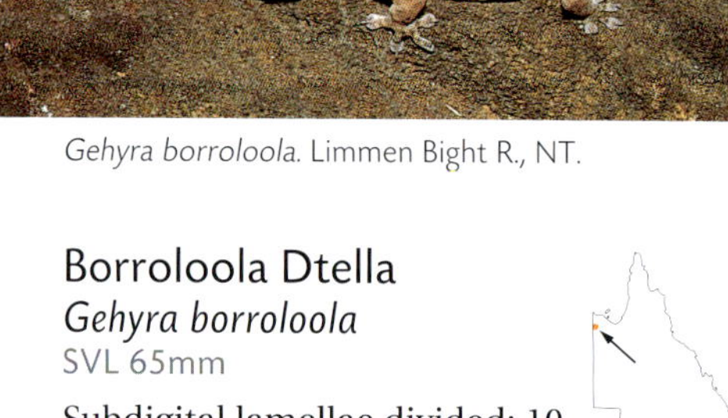

Gehyra dubia. Girraween NP.

Borroloola Dtella

Gehyra borroloola

SVL 65mm

Subdigital lamellae divided; 10 or more pairs under expanded pad of 4th toe; no web of skin at rear of hindlimb. Pale pink to dark brown with alternating transverse rows of dark blotches and cream spots.

NWH, in gorges and outcrops of Musselbrook sector of Lawn Hill NP. Also adj. NT. Lives under rock slabs and in rock crevices.

Dubious Dtella

Gehyra dubia

SVL 65mm

Subdigital lamellae undivided but sometimes deeply notched; 9 or more under expanded pad of 4th toe; no web of skin at rear of hindlimb; rostral scale margined above by 3–4 scales. Shades of grey to brown with irregular bands of pale spots, each dark-edged anteriorly. Dark markings often form netted pattern or pair of irregular longitudinal lines.

Widespread over vast tracts of timbered and rocky areas (with possible exception of CHC) in all but moistest regions. Also nthn interior of NSW. Shelters under loose bark, in rock crevices and human dwellings. Most abundant house gecko over much of its range.

Gehyra einasleighensis. Cobbold Gorge

Einasleigh Rock Dtella

Gehyra einasleighensis

SVL 40mm

Subdigital lamellae divided; 4–6 under expanded portion of 4th toe; no web of skin at rear of hindlimb; spotted pattern. Golden to dark tan with scattered prominent small circular pale spots and slightly larger, irregularly shaped dark blotches, sometimes with yellowish-orange edges.

Endemic. Rock-inhabiting among small boulders and rock rubble in EIU.

Amber Rock Dtella

Gehyra electrum

SVL 50mm

Subdigital lamellae undivided; 7–8 under expanded portion of 4th toe; no web of skin at rear of hindlimb; spotted pattern. Orange brown to pinkish orange typically with small white spots often arranged transversely,

S. Zozaya

Gehyra electrum. Springfield Stn.

and dark spots often merged into irregular blotches or bands.

Endemic. Granite outcrops and large boulders including vertical faces and overhangs in Mt Surprise area, EIU.

Gehyra lauta. Mt Isa.

Gulf Tree Dtella

Gehyra lauta

SVL 83mm

Subdigital lamellae undivided; 9–10 under expanded portion of 4th toe; no web of skin at rear of hindlimb; 22–32 preanal pores often extending onto limbs and reduced in size distally; very large and robust; weak pattern. Cream to pale silvery grey or purplish brown with little or no pattern on adults; indistinct darker bands and blotches on juv.

Arboreal, usually on smooth-barked trees and often along watercourses in rocky ranges of NWH. Also ne. NT.

Moritz's Dtella

Gehyra moritzi

SVL 49mm

Subdigital lamellae divided; 7–8 under expanded portion of 4th toe; no web of skin at rear of hindlimb; spotted pattern. Pinkish grey to reddish brown with small regular pale spots and larger irregularly shaped dark spots.

Rock-inhabiting, on ranges and outcrops at Craven's Peak, and probably other rocky habitats in the region, in far wstn CHC. Also sthn to central NT.

Gehyra moritzi. Kings Creek Stn, NT.

Purple Dtella

Gehyra purpurascens

SVL 64mm

Subdigital lamellae divided; 6–9 pairs under expanded pad of 4th toe; no web of skin at rear of hindlimb; top of rostral scale almost horizontal when viewed from above; no pale spots. Pale purplish grey with irregular darker spots or short streaks forming netted pattern.

CHC; widespread through arid zones to interior of WA. Arboreal, ranging from large trees to loose bark on sparse shrubs.

E. Vanderduys

Gehyra purpurascens. Craven's Peak.

Gehyra purpurascens. Windorah, Qld.

Gehyra robusta. Dajarra.

Robust Dtella

Gehyra robusta

SVL 75mm

Subdigital lamellae undivided; 9–12 under expanded pad of 4th toe; no web of skin at rear of hindlimb; relatively flat upper edge to rostral scale, margined above by only 2 large scales; relatively dorsally depressed body and long flat snout. Pinkish brown to orange-brown with bands of obscure pale spots alternating with transversely aligned dark blotches, often coalesced to form bars.

Endemic. Rocky ranges and outcrops in NWH; isolated weathered outcrops in wstn MGD. Shelters in narrow rock crevices; forages on exposed faces.

Variable Dtella

Gehyra versicolor

SVL 54mm

Subdigital lamellae divided; 7–8 pairs under expanded pad of 4th toe; no web of skin at rear of hind limb; upper edge of rostral scale normally steeply sloping. Greyish brown to grey with highly variable pattern; usually a dark network of irregular streaks, or transverse and longitudinal dark bars and irregular rows of pale dots, each often contacting the rear edge of a dark marking.

Vast tracts of Qld interior, wherever dry conditions prevail. Also interior of estn Aust. Abundant in any dry wooded country. Dwells mainly on trees and rocks, but also thrives under any ground debris and on walls of human dwellings.

Gehyra versicolor. Julia Creek.

Gehyra versicolor. Currawinya NP.

Genus *HEMIDACTYLUS*

About 75+ spp. worldwide; 1 in Aust. (introduced).

Hemidactylus frenatus. Kurwongbah.

Hemidactylus frenatus. Kurwongbah.

Asian House Gecko
Hemidactylus frenatus
SVL 60mm

Fine body scales mixed with small scattered tubercles; bands of larger tubercles on original tail; digits all clawed, flattened and expanded to form pads; subdigital lamellae expanded in paired series along digits; pupils have undulating margins. Grey with irregular dark stripes by day, ghostly pale and patternless at night.

Rapidly expanding urban and rural distribution. In early 1980s, stowaways occasionally recorded near Brisbane's international container terminals. By mid-1990s, abundance/distribution had grown exponentially. Now present in towns across Qld and nthn Aust. Rarely strays far from buildings, catching insects attracted to lights and readily betraying presence with distinctive 'chuck … chuck … chuck' call. World's most widespread lizard; introduced pops occur pantropically.

Genus *HETERONOTIA* 3 Aust. spp.; 1 occurs in Qld.

Bynoe's Gecko
Heteronotia binoei
SVL 54mm

Slender, padless, angular digits with strong claws; 2 small plate-like scales under base of each claw; fine scales mixed with numerous large, scattered to longitudinally aligned tubercles. Colour and pattern variable; typically brown to reddish brown with dark and pale markings forming irregular bands.

Heteronotia binoei. Morney Stn.

Heteronotia binoei. Lawn Hill NP.

Throughout Qld except most of SE; all mainland States. Favours dry open habitats; tends to shun closed moist areas. Terrestrial, normally very abundant, sheltering under any available ground debris. Much swifter than most ground geckos, fleeing rapidly when disturbed.

Genus *LEPIDODACTYLUS*

About 25 spp. from South Asia to Pacific; 2 occur in Qld. Digits flattened to form pads; subdigital lamellae expanded along length of digit, either all entire or some divided towards tip; all digits clawed except innermost, claw arising from the front edge of each pad; fine uniform body scales; weakly depressed heads and bodies; small size.

Arboreal geckos that thrive on walls of tropical, near-coastal dwellings.

Lepidodactylus lugubris. Cairns.

Mourning Gecko

Lepidodactylus lugubris

SVL 50mm

Original tail long, slightly flattened, edged with distinctive fine fringe of soft, spiny scales; last few subdigital lamellae under expanded portion of digit divided. Cream, pink to brown with a series of dark W-shapes, often reduced to few prominent, scattered spots.

Occurs mainly on dwellings in CYP, WT, CQC, s. to recently introduced pop. on Sunshine Coast. Also widespread pantropically. All-female sp. producing cloned offspring, no doubt contributing to its success as island colonist across wstn Pacific. One survivor on raft of dislodged vegtn or amongst human cargo is sufficient to establish new, viable population.

E. Rittmeyer

Lepidodactylus pumilus. Morehead area, PNG.

Slender Mourning Gecko

Lepidodactylus pumilis

SVL 48mm

Original tail long and cylindrical, with no fringed edges; all subdigital lamellae entire. Pale pinkish brown with narrow irregular dark bands on pale vertebral region.

Is. of Torres Strait incl. Murray, Hammond, Prince of Wales, and reported but not confirmed on CYP mainland s. to Jardine R. Also Daru, se. NG. Very poorly known. Recorded on walls of dwellings and in coastal forests.

Genus *NACTUS*

About 8 spp. from wstn Indian Ocean to wstn Pacific; 4 occur in Qld. Slender, padless digits, angular in profile with each claw set above a single curved plate-like scale; fine scales mixed with longitudinal rows of large tubercles; slender tail; small size.

Tropical woodlands and outcrops. Most occur abundantly under any available groundcover, sometimes in close proximity to the superficially very similar Bynoe's Gecko (*Heteronotia binoei*). One sp. a highly specialised boulder inhabitant with extremely restricted distribution. All are swift, alert geckos.

KEY TO *NACTUS*

1 Extremely slender with very long limbs and digits; large pale dorsal blotches or bands; Black Mountain only .... ***galgajuga***
Body and limbs moderate; dorsal pattern of small flecks 2

2(1) Subcaudal scales keeled........ ***'pelagicus'***
Subcaudal scales smooth 3

3(2) Preanal pores, if present, 1–5, usually 3 or fewer; thighs smooth (about 50 per cent of individuals); south of Cape Melville........ ***cheverti***
Preanal pores, if present, 4–11, usually 6 or more; thighs normally with tubercles; north of Princess Charlotte Bay........ ***eboracensis***

Southern Cape York Nactus
Nactus cheverti
SVL 57mm

One of 2 extremely similar spp. with multi-keeled, conical dorsal tubercles and smooth scales beneath tail, distinguishable only by a suite of discrete characteristics; dorsal tubercles (measured between rear edge of forelimb and front edge of hindlimb) in 15–27 (usually 20 or fewer) longitudinal rows; pores (small but sometimes present in front of vent of male) 1–5, usually 3; thighs smooth, without large tubercles on about half of individuals. Brown, usually with small pale blotches, each with darker leading edge. Markings become more continuous on tail, forming ragged bands.

Endemic. Nthn EIU, WT, sthn CYP, n. to Cape Melville.

Nactus cheverti. Shipton's Flat.

Northern Cape York Nactus
Nactus eboracensis
SVL 57mm

Like *N. cheverti*, has multi-keeled, conical dorsal tubercles and smooth scales beneath tail; dorsal tubercles in 15–37 (usually more than 22) longitudinal rows; pores (when present) 4–11, usually 6; thighs almost always have large tubercles.

Endemic. Nthn CYP, from Princess Charlotte Bay to sthn Torres Strait, incl. Thursday, Wednesday, York, Yam Is.

Nactus eboracensis. Tip of Cape York.

Black Mountain Gecko
Nactus galgajuga
SVL 50mm

Extremely slender; very large eyes; slightly upturned snout; long thin limbs and tail; smooth scales beneath tail. Dark purplish brown with broad, irregular pale grey bands, diffuse on body, becoming sharp-edged black and white on tail.

Endemic; one of the most restricted distributions of any Australian lizard, surreal piled granite boulders of Black Mtn, just s. of Cooktown, CYP. By day retreats into cavities between boulders, emerging at night to forage with great agility on exposed rock surfaces.

Nactus galgajuga. Black Mtn.

Pelagic Gecko
Nactus 'pelagicus'
SVL 57mm

Superficially very similar to *N. cheverti* and *N. eboracensis*, but readily distinguished by scales under tail. These are keeled, a feature common on *Nactus* elsewhere, but all other spp. endemic to Aust. have smooth subcaudal scales.

Some Torres Strait is. s. to Prince of Wales Is. Very similar, all-female pops occur in NG and many isolated is. of the Pacific. They reproduce by parthenogenesis, where unfertilised eggs produce clones of the mother. '*Pelagicus*' refers to its far-flung distribution across Oceania. Occupies various lowland forest types but appears to thrive best in disturbed regrowth and plantation habitats.

Nactus 'pelagicus'. Prince of Wales Is.

FLAP-FOOTED LIZARDS

Burton's Snake-lizards (*Lialis burtonis*) have a unique hinge across the head, ensuring a secure grip on their lizard prey. Badu Island.

Family **Pygopodidae**

Long-bodied, long-tailed lizards with lidless eyes covered by transparent spectacles; ear openings; no front limbs; hindlimbs reduced to scaly flaps. At first glance they appear limbless and snake-like. Unlike snakes, ventral scales are in a paired series, and original tails are much longer than the body (up to 4 times in some species). The tail is easily broken; a regenerating tail is normally visible by an abrupt change in scale arrangement and pattern.

The family includes diurnal and nocturnal species, dwelling in thick low vegetation such as grass tussocks and leaf litter. When active in open terrain they often move with a series of wriggling leaps. Their closest allies are geckos, sharing a voice, the ability to wipe the face with a broad flat tongue, and a fixed clutch size of 2 eggs. Most flap-footed lizards eat arthropods, but Burton's Snake-lizard (*Lialis burtonis*) is a reptile specialist.

KEY TO GENERA

a b

1	Snout sharp, acutely wedge-shaped	*Lialis*
	Snout rounded	2
2(1)	Preanal pores absent	*Delma*
	Preanal pores present	3
3(2)	Preanal pores 4 **(a)**	*Paradelma*
	Preanal pores 8 or more **(b)**	*Pygopus*

Genus *DELMA*

22 Aust. spp.; 9 occur in Qld. Well-developed hindlimb flaps; long tails up to 4 times length of body; no preanal pores; smooth shiny scales; ventral scales in paired series, noticeably larger than adj. scales.

Diversity greatest in arid zones featuring spinifex, but *Delma* occur statewide, mainly in dry grassy areas, coastal heaths, dry sclerophyll forests, and even rainforest margins. Diurnal and nocturnal, very secretive and infrequently encountered active. When disturbed they retreat rapidly through thick vegtn.

KEY TO *DELMA*

a b c d e f

1	Two preanal scales **(a)**	2
	Three preanal scales **(b)**	3
2(1)	Throat boldly marbled or reticulated	*torquata*
	Throat patternless	*plebeia*
3(1)	Third upper labial scale below eye; midbody scales normally 14	*tincta*
	Fourth upper labial scale below eye; midbody scales normally 16	4
4 (3)	Two narrow dark longitudinal stripes present	5
	No narrow dark stripes	6
5(4)	Ventrolateral stripe; banded head	*mitella*
	Dorsolateral stripe; face and sides of neck barred	*labialis*
6(4)	Some indication of dark bands across top of head	*borea*
	No indication of dark bands across top of head	7
7(6)	Two **(c)** or 4 **(d)** supranasal scales – if 4, then posterior pair clearly separated from nostril **(e)**	*inornata*
	Four supranasal scales – posterior pair contacting or narrowly separated from nostril **(f)**	8
8(7)	Snout long; dark spots on dorsal and ventral surfaces; face patternless	*nasuta*
	Snout short; no dark spots; face pattern (if present) pale vertical bars	*butleri*

Delma borea. Mt Isa district.

Northern Delma

Delma borea

SVL 98mm

Three preanal scales; 4 supranasal scales; midbody scales in 16 rows; 4th upper labial scale below eye; prominent dark bands across head and neck on juv. and young adult. Brown, reddish brown to grey with 3–4 dark bands on head and nape, separated by narrow pale interspaces tending to be orange dorsally, cream laterally.

CHC, NWH, probably wstn parts of MGD. Also NT, nthn WA. Associated with spinifex, mainly on stony soils.

Delma butleri. Nappa Merrie Stn.

Spinifex Delma

Delma butleri

SVL 96mm

Three preanal scales; 4 supranasal scales; midbody scales in 15–18 (usually 16) rows; 4th upper labial scale below eye. Brown, with fine dark edges to scales, series of narrow, wavy, pale vertical bars on sides of head and neck.

Spinifex on dunes, sandy interdunes in CHC, e.to Windorah area. Also arid parts of all mainland States.

Delma inornata. Bowenville.

Olive Delma

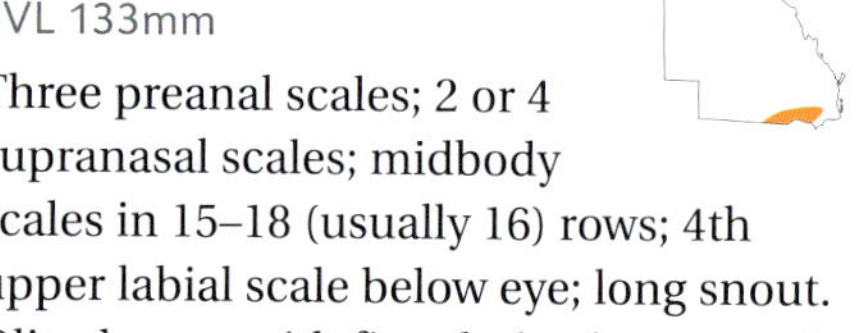

Delma inornata

SVL 133mm

Three preanal scales; 2 or 4 supranasal scales; midbody scales in 15–18 (usually 16) rows; 4th upper labial scale below eye; long snout. Olive brown with fine dark edges to scales, and cream to yellow below.

Sthn BB, in grasslands from Darling Downs w. to Culgoa area. Also dry temperate NSW, Vic, estn SA. One of 3 essentially sthn grassland species (others *Tympanocryptis pinguicolla*, *Carlia tetradactyla*) that reach their nthn limits in grasslands of Darling Downs, most of which now exists as narrow road verges.

Stripe-tailed Delma

Delma labialis

SVL 115mm

Three preanal scales; 4 supranasal scales; midbody scales in 16–18 rows; 4th upper labial scale below eye; long snout; barred neck; dark stripe on tail. Reddish brown with yellow flush on head, series of prominent vertical bars on sides of head, neck and forebody, narrow dark dorso-lateral stripe on tail and posterior body.

Delma labialis. Keswick Is.

Endemic. CQC, s. to Keswick Is. off Mackay, and far nthn BB from Magnetic Is. to foothills below Paluma. Isolated pop. at Heathlands , CYP. Normally associated with low open forest over thick grasses; also recorded in wet sclerophyll forest. Diurnal; extremely wary, usually glimpsed fleetingly.

Delma mitella. Tolga.

Atherton Delma

Delma mitella

SVL 154mm

Three preanal scales; 4 supranasal scales; midbody scales in 16 rows; 4th upper labial scale below eye; long snout, pale bands across head; dark bluish grey ventrolateral stripe; very large size. Reddish brown with 4 narrow pale bands across dark copper head (becoming indistinct to absent with age), narrow dark stripe along abrupt junction between flanks and bright yellow ventral surfaces.

Endemic. WT, adj. estn parts of EIU s. to Paluma district, in open forests and rainforest interfaces. Largest *Delma*, known from only a few specimens.

Apparent restriction to areas of moderately high rainfall contrasts with other members of this essentially dry-adapted group. Probably favours tussocks in well-drained habitats.

Delma nasuta. Windorah.

Sharp-snouted Delma

Delma nasuta

SVL 112mm

Three preanal scales; 4 supranasal scales; midbody scales in 16 rows; 4th upper labial scale below eye; long snout. Brown to olive with dark markings on each scale creating reticulated or simple spotted pattern. No dark markings on head or nape.

Spinifex, mainly on red sands and loams, in CHC, wstn MGD, NWH. Also arid zones of NT, SA, WA.

Leaden Delma

Delma plebeia

SVL 122mm

Two preanal scales; 4 supranasal scales; midbody scales in 14–16 (usually 16) rows; 4th upper labial scale below eye. Shades of grey or brown to olive with dark head markings reduced to small, diffuse blotches on face and sides of neck, often obscure, but normally discernible, even on aged individuals.

Delma plebeia. Pittsworth.

Heaths and dry forests of coastal SEQ, incl. disturbed areas adj. to Brisbane, w. to brigalow communities and sand-plains with spinifex in sthn BB. Also ne. NSW.

Black-necked Delma

Delma tincta

SVL 92mm

Three preanal scales; 2 supranasal scales; midbody scales in 14 rows; 3rd upper labial scale below eye; prominent dark head and neck markings on juv. and young adult. Shades of brown to reddish brown with 3–4 black bands, separated by cream to yellow interspaces on head and nape, fading on old adults.

Dry areas throughout Qld. Also NSW, SA, NT, WA.

Delma tincta. Sundown NP.

Collared Delma

Delma torquata

SVL 63mm

Two preanal scales; 2 supranasal scales; midbody scales in 16 rows; 3rd upper labial scale below eye; very prominent black head and neck markings extending well onto throat; very small size. Reddish brown, flushed with bluish grey on tail. Individuals of all ages have conspicuous shiny black bands across head and neck, broken by sharp, narrow, yellow interspaces. Chin and throat with bold grey to black reticulations or marbling.

Endemic. Recorded disjunctly from wstn edges of Brisbane in SEQ, nw. to Blackdown Tableland and inland to Roma area, BB. Mainly open rocky terrain though at least 2 sites feature woodlands of Forest River Gum (*Eucalyptus tereticornis*) or brigalow associations with no significant rock components. Smallest *Delma*.

Delma tincta. Winton.

Delma torquata. Mt Crosby.

Genus *LIALIS* 2 spp. in Aust. and NG; 1 occurs in Qld.

Burton's Snake-lizard

Lialis burtonis

SVL 290mm

Pointed, wedge-shaped snout; fragmented head shields; greatly reduced limb-flaps; preanal pores present; paired ventral scales usually noticeably wider than adj. body scales. Easily recognised by distinctive snout.

Lialis burtonis. Mt Tyson.

Bewildering array of colour forms: some appear influenced by environmental factors (high incidence of striped lizards in spinifex grasslands, pale specimens with broad black and white facial stripes in nthn woodlands) but others appear random. Colours range from pale grey to brown, rich red or yellow with pattern absent, comprising longitudinal rows of dark dashes, or prominent stripes.

Throughout Qld and all mainland States, wherever dry conditions prevail. Also NG (another sp., *L. jicari*, restricted to NG). Diurnal and nocturnal. Feeds exclusively on reptiles, mainly lizards. These are seized by ambush and grasped across chest until suffocated, completely enclosed by the long slender snout thanks to unusual hinge across head at about eye-level.

Genus *PARADELMA* Monotypic genus.

Brigalow Scaly-foot

Paradelma orientalis

SVL 197mm

Very glossy scales in 18–20 rows; ventrals in paired series, noticeably larger than adj. body scales; limb-flaps moderately large; 4 preanal pores; robust build. Greyish brown with shiny, milky sheen, black bar on nape, cream head darkening towards snout.

Endemic. BB, on sandstone ridges, dry forests and woodlands. Nocturnal. Primarily terrestrial but recorded climbing rough-barked wattles, possibly to lap exuding sap. When provoked, rears head and flickers tongue like small venomous snake.

Paradelma orientalis. Strathblane Stn.

Scaly-foots
Genus *PYGOPUS*

5 spp. in Aust.; all occur in Qld. Large, prominent, well-developed limb-flaps; 8 or more preanal pores; matt to weakly glossy body scales; ventral scales in paired series, noticeable larger than adj. body scales; large size.

Throughout Qld, wherever dry conditions prevail. Most are nocturnal though largest sp. (*P. lepidopodus*) is commonly encountered active by day. As members of a family characterised by extreme limb loss, their large limb-flaps suggest they are less highly modified than their relatives. When provoked scaly-foots rear their heads and flicker their tongues in apparent mimicry of small venomous snakes.

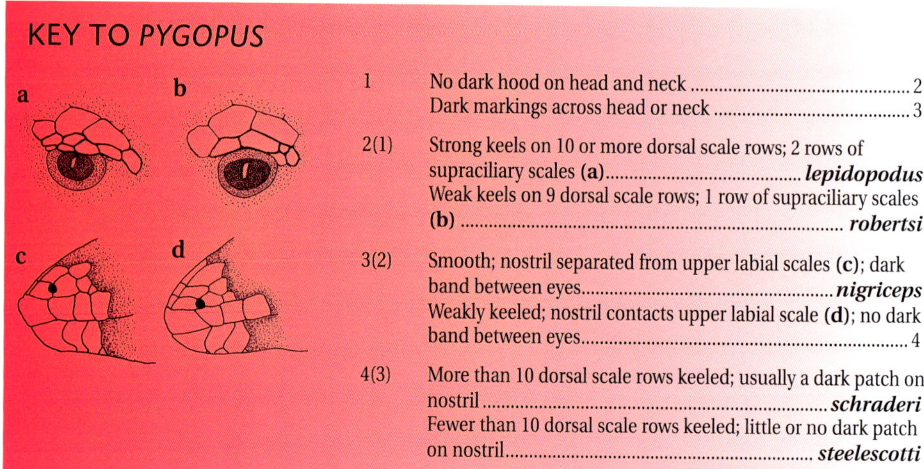

KEY TO *PYGOPUS*

1	No dark hood on head and neck	2
	Dark markings across head or neck	3
2(1)	Strong keels on 10 or more dorsal scale rows; 2 rows of supraciliary scales **(a)**	***lepidopodus***
	Weak keels on 9 dorsal scale rows; 1 row of supraciliary scales **(b)**	***robertsi***
3(2)	Smooth; nostril separated from upper labial scales **(c)**; dark band between eyes	***nigriceps***
	Weakly keeled; nostril contacts upper labial scale **(d)**; no dark band between eyes	4
4(3)	More than 10 dorsal scale rows keeled; usually a dark patch on nostril	***schraderi***
	Fewer than 10 dorsal scale rows keeled; little or no dark patch on nostril	***steelescotti***

Common Scaly-foot
Pygopus lepidopodus
SVL 274mm

Very strong keels on more than 10 dorsal and lateral scale rows; 2 rows of supraciliary scales; no dark markings on head and neck; very large size. Reddish-brown with weak pattern; sometimes a dark dash under eye and obscure longitudinal rows of dark dashes on tail. These markings seem to be relicts from much more boldly striped and spotted pops occupying sthn Aust. heaths.

Coast and ranges of SEQ to sthn CQC and BB, with outlying pop. at Chesterton Ra. NP, in ML. Also across southern Aust. Normally encountered amongst tussock grasses in dry open forests. Squeaks

Pygopus lepidopodus. Mt Nebo.

loudly if grasped, rotating within the hand and readily severing tail. Possibly Aust.'s largest pygopodid; large adult with complete tail can reach 1m. Diurnal to nocturnal.

Pygopus nigriceps. Nappa Merrie Stn.

Western Hooded Scaly-foot

Pygopus nigriceps

SVL 227mm

Smooth, weakly glossed scales; dark markings on head and neck; nostril not contacting 1st upper labial scale; usually 120 or more ventral scales; fewer than 14 preanal pores. Brown to reddish brown with broad band across head through eyes onto lips, dark smudge on nostril, broad dark band across neck. Little or no other pattern; obscure, oblique dark lines converging mid-dorsally, largely confined to tail.

Sandy deserts vegetated with spinifex in CHC. Also deserts of NT, SA, WA. Shelters in soil cracks, insect holes, under debris. Nocturnal.

Roberts' Scaly-foot

Pygopus robertsi

SVL 224mm

Weak keels on 9 dorsal scale rows; 1 row of supraciliary scales; no dark markings on head and

Pygopus robertsi (holotype). Little Forks, Shipton's Flat.

neck. Reddish brown to greyish brown with little pattern; dark bar below eye and sometimes faint grey chin straps.

Endemic. CYP and wstn edge of WT, from Yungaburra on Atherton Tableland n. to Shipton's Flat, with possibly isolated pop. in Heathlands area, nthn CYP. Heath and dry woodlands.

Eastern Hooded Scaly-foot

Pygopus schraderi

SVL 198mm

Scales keeled on 10 or more dorsal body rows; dark markings on head and neck; nostril in contact with 1st upper labial scale;

Pygopus schraderi. South Galway Stn.

fewer than 120 ventral scales; 13 or more preanal pores. Brown to reddish brown or grey with teardrop-shaped marks under eye and nostril, dark band of varying intensity across neck, no dark band between eyes. Pattern on body and tail prominent to absent, comprising diffuse dark and pale lines tending to converge on vertebral line, particularly on tail, and sometimes excluding an area on midline, creating pale vertebral stripe.

Widespread in dry woodlands and other open habitats, often on stony and heavy soils. Also dry parts of NSW, Vic, SA. Shelters in soil cracks, insect holes, under debris. Nocturnal.

Northern Hooded Scaly-foot

Pygopus steelescotti

SVL 185mm

Scales weakly keeled on fewer than 10 dorsal body rows; weak dark markings on head and neck; nostril in contact with 1st upper labial scale; fewer than 120 ventral scales; 13 or more preanal pores. Pale brown to pale yellowish brown with obscure pattern incl. dark marking under eye and dark band across neck, but little or no patch under nostril, no dark band between eyes. Body pattern weak to absent; dark and pale longitudinal lines tending to converge on vertebral line, particularly on tail.

Tropical woodlands of CYP, GUP, NWH. Also nthn NT and WA. Nocturnal.

C. Dollery

Pygopus steelescotti. Mt Isa.

SKINKS

This Yellow-blotched Forest-skink, (*Concinnia tigrina*), surveys its surroundings from a cavity in a rainforest log at Lake Barrine.

Family **Scincidae**

This large family is so diverse that only discrete internal features are shared by all members. The most significant are small bony pieces called osteoderms arranged as a symmetrical mosaic within each scale, and a complete or near-complete bony palate. Many skinks are so modified as to defy many of the external characteristics generally employed to define the group. Most have large symmetrical head shields, smooth overlapping body scales and 4 well-developed pentadactyl limbs.

Some burrowing skinks have developed long bodies and reduced or lost limbs and digits, with virtually all stages through to complete limblessness present within Qld. Those that forage over vertical rock faces and tree trunks are flat-bodied with long slender limbs and digits. Some inhabitants of crevices in logs or rocks have developed spiny scales, rendering them almost impossible for predators to extract. Some are large and ponderously slow. Others are lithe, swift and smaller than many insects.

Skinks range from insectivores to omnivores, and include livebearers and egglayers.

KEY TO GENERA

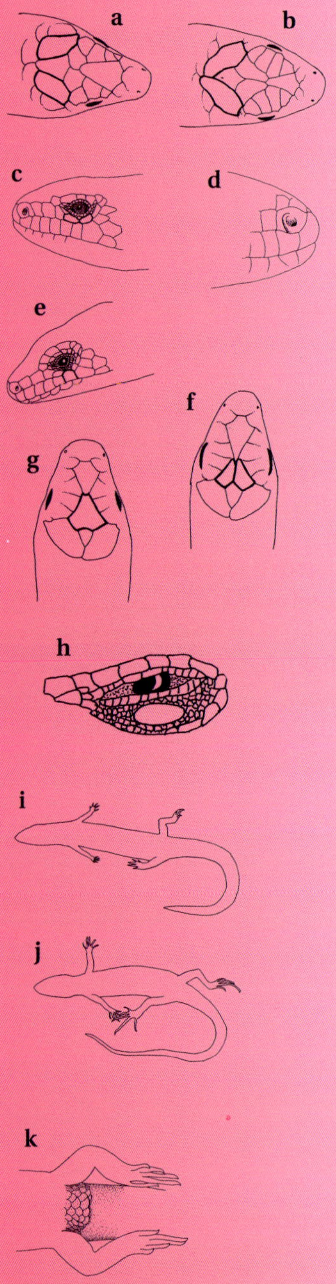

1 Parietal scales (sometimes fragmented) separated behind interparietal scale **(a)** 2
Parietal scales in contact behind interparietal scale **(b)** 7

2(1) Fourth toe much longer than 3rd........ 3
Fourth and 3rd toes approximately equal 6

3(2) No complete subocular scale row between eye and upper labials........ 4
A complete subocular scale row between eye and upper labials **(c)** .... ***Nangura***

4(3) Postnarial groove present, from nostril to 1st upper labial **(d)** 5
No postnarial groove .... ***Liopholis***

5(4) Eyelids usually with distinct cream margins, delineated from adjacent scales........ ***Bellatorias***
Eyelids without distinct cream margins, not delineated from adjacent scales........ ***Egernia***

6(2) A row of large subocular scales between eye and upper labials **(c)** .... ***Tiliqua***
No row of subocular scales, allowing upper labials to contact eye or granules below eye **(e)** .... ***Cyclodomorphus***

7(1) All scales, including ventrals, very strongly keeled .... ***Gnypetoscincus***
Scales on back and tail smooth to weakly keeled but all ventrals always smooth........ 8

8(7) Fingers 5 9
Fingers fewer than 5........ 29

9(8) Frontoparietal scales divided **(f)** 10
Frontoparietal scales fused to form one shield **(g)**........ 23

10(9) Lower eyelid fused to form a fixed, unblinking spectacle........ ***Proablepharus***
Lower eyelid movable (able to blink)........ 11

11(10) Lower eyelid encloses a transparent disc **(h)**........ 12
Lower eyelid scaly, with no transparent disc........ 14

12(11) Pale spot in posterior base of thigh ...... ***Saproscincus*** (part)
No pale spot in posterior base of thigh (or if present, part of more extensive lateral spotting) 13

13(12) Limbs short, failing to overlap when pressed to side of body **(i)**; southern uplands .... ***Harrisoniascincus***
Limbs long, easily overlap when pressed to side of body **(j)**; northern uplands .... ***Techmarscincus***

14(11) Rows of low longitudinal ridges on rump and base of tail........ ***Eremiascincus*** (part)
No ridges on rump and base of tail 15

15(14) Preanal scales not greatly enlarged, arranged in a row roughly equal to adjacent scales **(k)** .... ***Eugongylus***
2–3 greatly enlarged median preanal scales, much larger than adjacent scales **(l)** 16

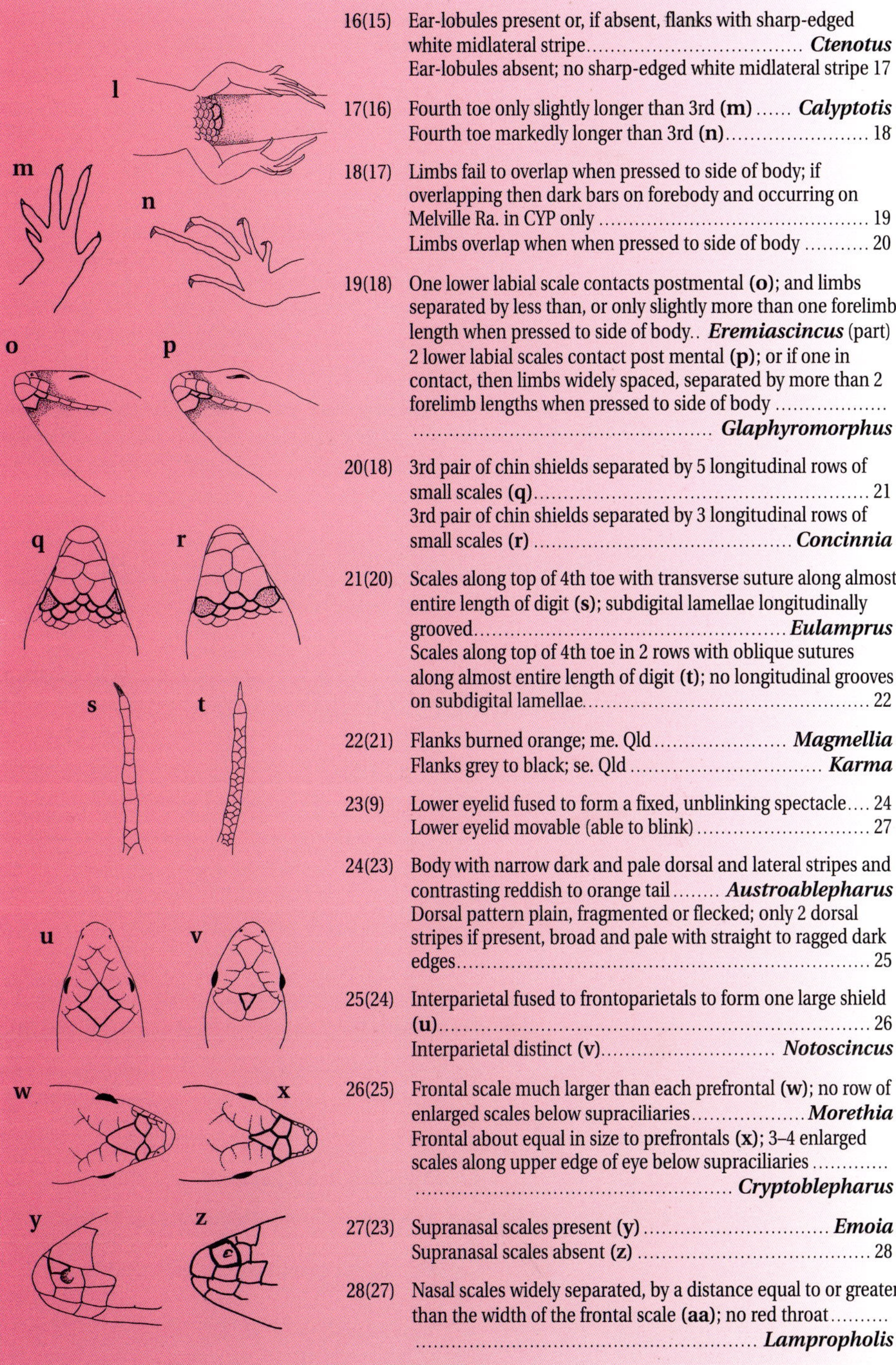

16(15) Ear-lobules present or, if absent, flanks with sharp-edged white midlateral stripe ... ***Ctenotus***
Ear-lobules absent; no sharp-edged white midlateral stripe 17

17(16) Fourth toe only slightly longer than 3rd (**m**) ... ***Calyptotis***
Fourth toe markedly longer than 3rd (**n**) ... 18

18(17) Limbs fail to overlap when pressed to side of body; if overlapping then dark bars on forebody and occurring on Melville Ra. in CYP only ... 19
Limbs overlap when when pressed to side of body ... 20

19(18) One lower labial scale contacts postmental (**o**); and limbs separated by less than, or only slightly more than one forelimb length when pressed to side of body.. ***Eremiascincus*** (part)
2 lower labial scales contact post mental (**p**); or if one in contact, then limbs widely spaced, separated by more than 2 forelimb lengths when pressed to side of body ... ***Glaphyromorphus***

20(18) 3rd pair of chin shields separated by 5 longitudinal rows of small scales (**q**) ... 21
3rd pair of chin shields separated by 3 longitudinal rows of small scales (**r**) ... ***Concinnia***

21(20) Scales along top of 4th toe with transverse suture along almost entire length of digit (**s**); subdigital lamellae longitudinally grooved ... ***Eulamprus***
Scales along top of 4th toe in 2 rows with oblique sutures along almost entire length of digit (**t**); no longitudinal grooves on subdigital lamellae ... 22

22(21) Flanks burned orange; me. Qld ... ***Magmellia***
Flanks grey to black; se. Qld ... ***Karma***

23(9) Lower eyelid fused to form a fixed, unblinking spectacle.... 24
Lower eyelid movable (able to blink) ... 27

24(23) Body with narrow dark and pale dorsal and lateral stripes and contrasting reddish to orange tail ... ***Austroablepharus***
Dorsal pattern plain, fragmented or flecked; only 2 dorsal stripes if present, broad and pale with straight to ragged dark edges ... 25

25(24) Interparietal fused to frontoparietals to form one large shield (**u**) ... 26
Interparietal distinct (**v**) ... ***Notoscincus***

26(25) Frontal scale much larger than each prefrontal (**w**); no row of enlarged scales below supraciliaries ... ***Morethia***
Frontal about equal in size to prefrontals (**x**); 3–4 enlarged scales along upper edge of eye below supraciliaries ... ***Cryptoblepharus***

27(23) Supranasal scales present (**y**) ... ***Emoia***
Supranasal scales absent (**z**) ... 28

28(27) Nasal scales widely separated, by a distance equal to or greater than the width of the frontal scale (**aa**); no red throat ... ***Lampropholis***

aa bb

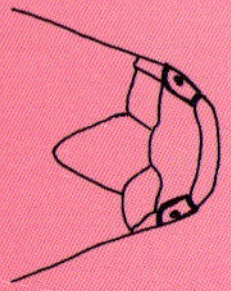

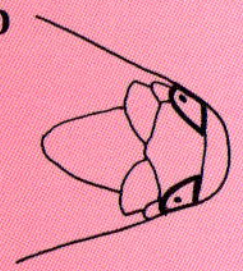

f g

Nasal scales moderately narrowly separated, by a distance less than the width of the frontal scale **(bb)**; usually a red throat. .... ***Acritoscincus***

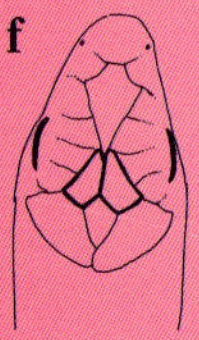

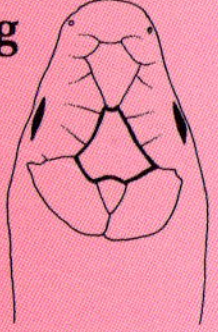

29(8) Fingers 4 and toes 5 30
Fingers and toes 4 or fewer 36

30(29) Frontoparietal scales divided **(f)** 31
Frontoparietals fused to form one shield **(g)** 32

31(30) A pale spot in rear base of hind limb; north-east .... ***Saproscincus*** (part)
No pale spot in rear base of hind limb; south-east .... ***Eroticoscincus***

32(30) Ear-opening conspicuous with lobules 34
Ear-opening minute without lobules 33

33(32) Interparietal scale fused to frontoparietals to form one large shield **(cc)** .... ***Pygmaeascincus***
Interparietal scale distinct **(dd)** .... ***Menetia***

34(32) Ten or more scales along top of 4th toe; pattern variable but never simple, straight-edged pale stripes ... 35
Usually fewer than 10 scales along top of 4th toe; if more, then straight-edged pale stripes present .... ***Lygisaurus***

35 (34) Dorsal scales smooth or with keels; if keeled, and more than 34 midbody scale rows, then no series of raised points along keels, and ear-lobules squared or rounded and confined to anterior margin .... ***Carlia***
Dorsal scales keeled with a series of raised points along keels; if evenly keeled, then 34 or more midbody scale rows and acute lobules around ear opening .... ***Liburnascincus***

36(29) Ear-opening very small but distinct .... ***Lerista***
Ear-opening hidden, represented at most by only a scaly depression 37

37(36) Snout wedge-shaped in profile; Fraser Island only .... ***Coggeria***
Snout round in profile 38

cc dd

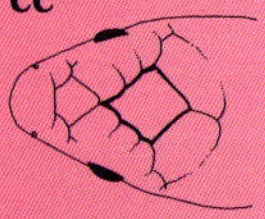

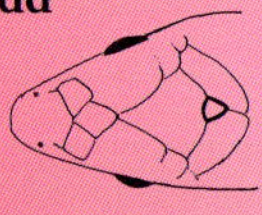

ee ff

gg hh

38(37) Limbs present each with 3 digits; 22 or fewer midbody scale rows .... ***Saiphos***
Limbs present or absent; digits variable if present, but if all limbs have 3 digits, then midbody scales in 23 or more rows .. 39

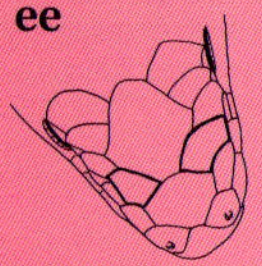

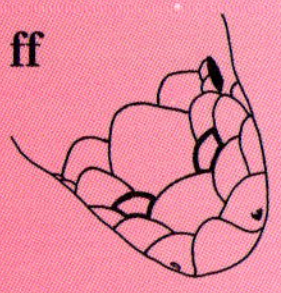

39(38) Limbs, if present, each with 3 digits; if absent, prefrontal scales moderately large and narrowly to moderately separated **(ee)** .... ***Coeranoscincus***
Limbs, if present, with no more than 2 digits on hind limb; if absent, prefrontal scales very small and widely separated **(ff)** or absent 40

40(39) Limbs present, with 2 or more fingers .... ***Anomalopus***
Limbs absent, or all reduced to minute stumps 41

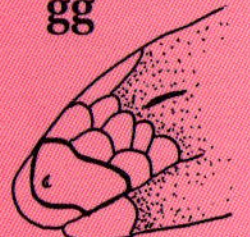

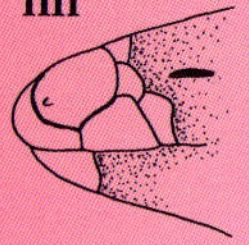

41(40) Orange/yellow ventral pigment .... ***Ophioscincus***
No orange/yellow ventral pigment 42

42(41) Nasal scale fused to 1st upper labial scale; ne CYP **(gg)** .... ***Suppressascincus***
Nasal scale distinct from 1st upper labial scale **(hh)** .... ***Praeteropus***

Note. Some of these genera include species occurring outside of Queensland which exhibit characters that would not apply to this key.

Genus *ACRITOSCINCUS* 3 Aust. spp.; 1 occurs in Qld.

Red-throated Skink
Acritoscincus platynotum
SVL 80mm

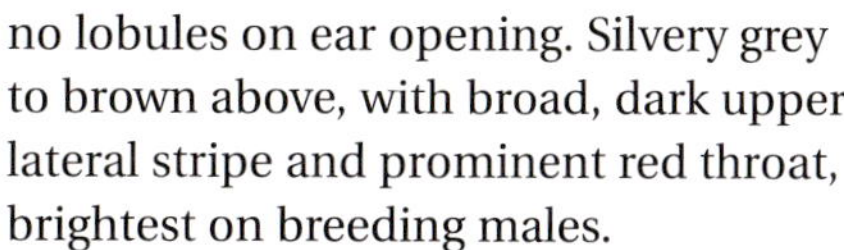

Red throat; movable lower eyelid enclosing transparent disc; parietal scales in contact; fronto-parietal scales fused to form one shield; interparietal scale small; nasal scales moderately widely separated; short, well-developed limbs each with 5 digits; no lobules on ear opening. Silvery grey to brown above, with broad, dark upper lateral stripe and prominent red throat, brightest on breeding males.

NET. Also estn NSW, estn Vic. Cool temperate sp.; close allies confined to sthn areas. Diurnal, foraging and basking amongst leaf litter and low vegtn, sheltering under rocks, logs and debris. Egglayer.

Acritoscincus platynotum. Girraween NP.

Genus *ANOMALOPUS*

4 Aust. spp.; 3 in Qld. Long bodied with limbs and digits greatly reduced; 3 or fewer fingers uniquely outnumbering toes; body scales smooth and glossy; parietal scales in contact; lower eyelid scaly with no transparent disc; ear opening represented by a scaly depression; prefrontal scales small and widely separated; nasal scales separated; snout round, not pointed or wedge-shaped.

Mid-eastern to south-eastern Qld, mainly in humid shelters associated with dry sclerophyll forests, heaths and stony slopes. Burrowing skinks inhabiting upper layers of soft earth beneath rocks, logs or leaf litter. All Qld spp. egglayers.

KEY TO *ANOMALOPUS*

1	Two fingers	***leuckartii***
	Three fingers	2
2	Hind limb a clawless stump	***verreauxii***
	Hind limb with 2 toes	***mackayi***

Anomalopus leuckartii. Barakula SF.

Two-clawed Worm-skink

Anomalopus leuckartii

SVL 137mm

2 fingers and hindlimb reduced to clawless stump. Brown to purplish brown with obscure pale longitudinal lines. N. of NET pale bar often extends across nape, but in NET nape is undifferentiated, at least on adults.

NET, SEQ, BB. Also ne. NSW. Woodlands, dry sclerophyll forests. Where it occurs near *A. verreauxii* (e.g. on Darling Downs), tends to favour drier, better drained sites.

Anomalopus mackayi. Bowenville.

Five-clawed Worm-skink

Anomalopus mackayi

SVL 123mm

3 fingers and 2 toes. Greyish brown with longitudinal rows of dark spots, one per scale, over dorsal, lateral and sometimes ventral surfaces.

Sthn BB. Also ne. interior of NSW. Open grasslands on heavy cracking soil in Darling Downs. Normally found in soil under dead grass. Largely confined to relict roadside verges.

Anomalopus verreauxii. Coominya.

Three-clawed Worm-skink; Verreaux's Skink

Anomalopus verreauxii

SVL 185mm

3 fingers and hindlimb reduced to clawless stump. Brown to grey with broad pale bar across base of head. On juv. this bar prominent and yellow; on adults much darker but usually discernible.

SEQ, BB, CQC n. to about Proserpine. Also ne. NSW. Dry sclerophyll forests, woodlands, clearings in wet sclerophyll forests. Very tolerant of disturbed habitats, thriving in compost heaps in suburban gardens, beneath logs in cleared paddocks.

Grassland Striped Skinks
Genus *AUSTROABLEPHARUS*

3 Aust spp. 2 in Qld. Small and slender with well-developed limbs each with 5 digits; lower eyelid fused to form a fixed spectacle; ear-opening very small; frontoparietal scales fused to form one shield, interparietal free or fused to frontoparietals, parietal scales in contact. Pattern comprises stripes, formed by dark lateral edges to scales, and a red tail.

Arid to semiarid grasslands on cracking clay in interior of Qld. Also NT and NSW. Diurnal but extremely cryptic, seldom venturing into open areas and sheltering in soil cracks and thick low vegtn. Egglaying.

Austroablepharus barrylyoni. Springfield Stn.

Austroablepharus kinghorni. Durham Downs Stn.

Lyon's Grassland Striped Skink
Austroablepharus barrylyoni
SVL 51mm

Interparietal and frontoparietal scales fused to from one shield. Dark brown with 10–14 narrow pale stripes; prominent dorsals, dorsolaterals and midlaterals from behind head to proximal third of tail, and ill-defined, variable lateral stripes. Breeding males have orange-red flush on outer edge of throat and jaw. Straw-coloured tails of preserved specimens suggest orange flush in life.

Endemic. Grassland with scattered trees on seasonally inundated cracking clay soils at Springfield Stn, EIU. Sole known locality severely impacted by cattle since specimens collected. No live image available.

Kinghorn's Grassland Striped Skink
Austroablepharus kinghorni
SVL 45mm

Interparietal scale free from frontoparietals. Pale brown to almost cream with stripes conspicuous and extending along full body on most specimens, but obscure and mainly restricted to forebody on large adults. Tail flushed with red.

Arid grasslands on deeply cracking clay soils, heavy red loams of CHC, ML, MGD. Also NSW, NT. Shelters in deep soil cracks, beneath overhanging skirts of tussocks, under leaf litter.

Genus *BELLATORIAS*

3 Aust. spp.; 2 occur in Qld. Extremely large and robust; 4 well-developed limbs, each with 5 digits; 4th toe much longer than 3rd; moveable lower eyelid, lacking a transparent disc; distinctive cream margins around eyelids; anterior lobules on ear-opening; parietal scales separated; smooth to weakly keeled body scales.

Rainforests, wet scleroplyll forests and other well-vegetated areas in estn Qld. Terrestrial, sheltering in burrows, hollow logs and crevices. Diurnal omnivorous livebearers. Gregarious, often living in family groups.

Major Skink

Bellatorias frerei

SVL 180mm

Extremely large size; low blunt dorsal keels. Brown with narrow dark lines above, darker brown with pale spots on flanks.

Forests, heaths, outcrops of SEQ, BB, CQC, EIU, WT, CYP, n. through Torres Strait to sthn NG. Also ne. NSW. Tends to be less common in sthn rainforests, where it is replaced by the Land Mullet (*B. major*). Usually very shy, basking in semi-concealed sites near the shelter of hollow logs, rocks and burrows. Pop. densities appear to be high on some offshore islands (e.g. Keswick).

Bellatorias frerei. Mt Glorious.

Land Mullet

Bellatorias major

SVL 300mm

Extremely large size; glossy black patternless adults; low blunt dorsal keels. Adults marked only with prominent pale rim around eye. Juv. has prominent cream lateral spots.

SEQ. Subtropical rainforests and adj. wet sclerophyll forests on ranges n. to Conondales. Also ne. NSW. Normally very shy, basking in sunny patches and dashing noisily to cover of dense low vegtn if approached. In some popular NPs, lizards become habituated and approachable, scavenging for scraps at picnic and camping sites. Shelters in hollow logs or burrows, often dug into soil-bound root systems of fallen trees. Like other large skinks, vegtn features prominently in diet, fungi being particularly favoured.

Bellatorias major. Adult and juvenile. Springbrook.

Genus *CALYPTOTIS*

5 Aust. spp.; 4 occur in Qld. Moderately long-bodied with very short limbs, each with 5 digits of about equal length; movable lower eyelid with no transparent disc; parietal scales in contact; frontoparietals paired; ear opening present or absent; most have yellow on chest and belly, pink under tail.

Estn Qld. Also ne. NSW. These skinks shelter in humid sites under rocks, logs and leaf litter in rainforests, wet and dry sclerophyll forests, suburban gardens. Egglayers.

KEY TO *CALYPTOTIS*

1	Ear opening present, exposing tympanum........	***temporalis***
	Ear opening absent..	2
2(1)	Prefrontal scales absent **(a)**	***scutirostrum***
	Prefrontal scales present **(b)**	3
3(2)	Ear represented by scaly conical depression; SEQ...............	***lepidorostrum***
	Ear represented by flat scaly disc; WT..........	***thorntonensis***

Cone-eared Calyptotis

Calyptotis lepidorostrum

SVL 55mm

No ear openings, only conical scaly depressions; prefrontal scales present; yellow and pink ventral pigments present. Brown above, usually with 4 narrow dark dorsal lines (at least from nape to forebody), broader dark dorsolateral stripe.

Endemic. Rainforests, wet and dry sclerophyll forests and heaths in SEQ, coastal BB, CQC.

Calyptotis lepidorostrum. Cooloola NP.

Garden Calyptotis

Calyptotis scutirostrum

SVL 55mm

No ear openings, only conical scaly depressions; no prefrontal scales; yellow and pink ventral pigments present. Brown above, usually with 4 narrow dark dorsal lines (at least from nape to forebody), broader dark dorsolateral stripe.

Rainforests, wet and dry sclerophyll forests, heaths, suburban gardens in SEQ, NET, BB, with isolated inland pop. in Mt Moffatt sector of Carnarvon NP. Also ne. NSW.

Calyptotis scutirostrum. Mt Glorious.

Broad-templed Calyptotis
Calyptotis temporalis
SVL 35mm

Large, obvious ear openings; no prefrontal scales; yellow and pink ventral pigments present. Brown above, usually with 4 narrow dark dorsal lines (at least from nape to forebody), broader dark dorsolateral stripe.

Endemic. Rainforests, wet and dry sclerophyll forests and heaths in coastal BB, CQC, n. to about Proserpine area.

Calyptotis temporalis. Miriam Vale.

Thornton Peak Calyptotis
Calyptotis thorntonensis
SVL 35mm

Ear openings absent, represented by flat scaly discs; no prefrontal scales; no yellow and pink ventral pigments; slightly flattened head and body. Brown above, with dark mottling, dark dorsolateral stripe.

Endemic. High altitude rainforests on steep sloping terrain at Roaring Meg and between 600 and 800m on Thornton Peak, WT. Moist environments under small rocks; reported to be so sensitive to heat that even limited handling can be lethal.

Calyptotis thorntonensis. Thornton Peak.

Four-fingered Skinks; Rainbow Skinks
Genus *CARLIA*

26 spp. from Aust., others through NG to estn Indonesia; 20 occur in Qld. Well-developed limbs with 4 fingers and 5 toes; moveable lower eyelid enclosing a transparent disc; conspicuous ear-opening with lobules; parietal scales in contact; frontoparietal scales fused to form one shield; smooth to strongly keeled dorsal scales.

Nthn and estn Qld, with centre of diversity in ne. Swift, sun-loving, strongly visually-cued leaflitter-inhabitants, normally featuring bold, spectacular breeding hues on mature males. Most exhibit tail-waving displays. Egglayers, producing 2 eggs per clutch.

KEY TO *CARLIA*

1 Interparietal scale fused with frontoparietal **(a)** 2
Interparietal scale free **(b)** 5

2(1) Ear-opening round to horizontal with sharp triangular anterior lobule, similar lobules around top and sometimes entire margin of ear; Melville Ra. in CYP only . ***wundalthini***
Ear-opening vertical to round with 1–3 large triangular anterior lobules and no or smaller lobules on other margins; WT to CQC........ 3

3(2) Throat and chin pink in life........ 4
Throat and chin pink and blue in life........ ***rhomboidalis***

4(3) Nthn WT, n. of an approximate line from Mareeba, through Lake Tinnaroo, crest of Lamb Range to sthn Cairns and e. of a line from sthn Cairns to mouth of Russell River .... ***crypta***
Sthn WT, s. of above approximate line........ ***rubrigularis***

5(1) Dorsal scales smooth to striated or very weakly-keeled; with smoothly curved hind edges **(c)**........ 6
Dorsal scales moderate to strongly-keeled; mostly with angular hind edges **(d)** 11

6(5) A white line from under eye to top of ear-opening, continuing back from bottom of ear-opening; transparent disc occupies much more than half lower eyelid **(e)**........ ***munda***
White line absent or not as above; transparent disc occupies about half of lower eyelid or less **(f)** (if larger, then restricted to ne. Torres Strait)........ 7

7(6) Ear-opening round with a blunt anterior lobule........ ***tetradactyla***
Ear-opening round to vertical with pointed lobules........ 8

8(7) Longitudinally aligned, dark-edged pale dashes on dorsum; ne. Torres Strait .... ***quinquecarinata***
No longitudinally aligned, dark-edged pale dashes on dorsum; Torres Strait to base of CYP 9

9(8) 1–2 large pointed anterior ear lobules; breeding male with black throat, dark speckling on back, black upper lateral stripe reaching nearly to hindlimb and red lower lateral stripe; female with prominent pale anterior dorsolateral and midlateral stripes reaching back to behind forelimb; base of CYP .... ***rostralis***
Usually at least several large pointed anterior ear lobules; breeding male with pale throat; back plain and flanks brown (flushed with red in breeding season); base of CYP to Torres Strait 10

10(9) Ear-opening completely surrounded by sharp pointed lobules; mature males have sharp contrast between pale lower neck and dark midlateral zone between ear and shoulder........ ***longipes***
Ear-opening with well-developed, comb-like anterior lobules, and lobules poorly to moderately developed in remaining margins; no sharp contrast between dark midlateral zone and lower neck .... ***sexdentata***

11(5) Dorsal scales with 3 keels........ 12
Dorsal scales with 2 keels........ 16

12(11) Usually 5 supraciliary scales........ 13
Usually 6 or more supraciliary scales........ 15

13(12) Upper preocular scale broadly triangular **(g)** or contacting posterior edge of 2nd loreal scale .... ***decora***
Upper preocular scale minute or a narrow vertical sliver **(h)**.. 14

14(13) Ear-opening round with prominent rounded or bluntly pointed lobules on all margins .... ***pectoralis***
Ear-opening usually vertically elliptic, with one low rounded anterior lobule, and sometimes low lobules on other margins .... ***rubigo***

i j k

15(12) Ear-opening with 2 large square anterior lobules **(i)** .... ***schmeltzii*** (part)
Ear-opening with a small pointed anterior lobule and small lobules on remaining margins **(j)** .... ***jarnoldae***

16(11) Ear-opening surrounded by pointed lobules **(k)** .... ***storri***
At least 1–2 larger rounded anterior ear lobules 17

17(16) Ear-opening round to horizontally elongate; nw. Qld .. ***amax***
Ear-opening vertical; estn Qld 18

18(17) Supraciliary scales 7 .... ***schmeltzii*** (part)
Supraciliary scales 5 19

19(18) Black patch on ear; usually 29 or more lamellae under 4th toe; pale vertebral stripe on females and juveniles; sandy e. coastal CYP only .... ***dogare***
No black patch on ear; fewer than 29 lamellae under 4th toe; no pale vertebral stripe on females and juveniles 20

20(19) Heavy black dorsal blotches, or up to 11 black dorsal stripes (some males); Whitsunday Island group only. .... ***inconnexa***
No heavy black dorsal blotches or black dorsal stripes .... ***vivax***

Two-spined Rainbow Skink
Carlia amax
SVL 40mm

Two strong keels on each hexagonal dorsal scale; large disc in lower eyelid; round to horizontally elongate ear opening, usually with small anterior lobule; robust build. Olive brown, usually with mix of small pale and dark spots, pale streak under eye, sometimes reaching back for varying distances to forelimb. Head copper on breeding males.

Leaf litter in dry rocky areas of NWH, wstn GUP. Also nthn NT, WA.

Carlia amax. Cloncurry.

Northern Wet Tropics Red-throated Rainbow Skink
Carlia crypta
SVL 54mm

Smooth dorsal scales with curved hind edges; interparietal scale fused to frontoparietal scale; round ear-opening with 1–2 large pointed anterior lobules and smaller pointed lobules on remaining edges; red throat. Rich brown with black flecks mainly on paravertebral zone, a narrow pale dorsolateral stripe, dark upper flanks, an indication of pale midlateral stripe

Carlia crypta. Thornton Peak, Qld.

and copper on top of head. Pink on chin and throat, brightest on breeding male. Distinguishable from *C. rubrigularis* only by distribution and genetics.

Endemic. Rainforests and margins in nthn WT, n. to Big Tableland and s to a line appr. from Mareeba, through Lake Tinaroo, crest of Lamb Ra. to sthn Cairns, and e. of a line from sthn Cairns to mouth of Russell R.

Carlia decora. Breeding male. Pallarenda.

Carlia decora. Female. Alligator Creek, Mt Elliot.

Elegant Rainbow Skink

Carlia decora

SVL 49mm

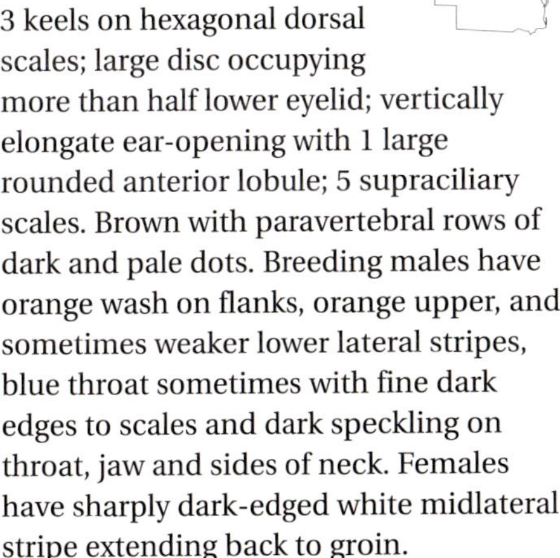

3 keels on hexagonal dorsal scales; large disc occupying more than half lower eyelid; vertically elongate ear-opening with 1 large rounded anterior lobule; 5 supraciliary scales. Brown with paravertebral rows of dark and pale dots. Breeding males have orange wash on flanks, orange upper, and sometimes weaker lower lateral stripes, blue throat sometimes with fine dark edges to scales and dark speckling on throat, jaw and sides of neck. Females have sharply dark-edged white midlateral stripe extending back to groin.

Endemic. Moist forests, vine thickets and RF margins in CQC, nthn BB and WT, between about Sarina and Mt Molloy.

Carlia dogare. Female. Cape Flattery.

Carlia dogare. Breeding male. Cape Flattery.

Sandy Rainbow Skink

Carlia dogare

SVL 50mm

Two strong keels on each hexagonal dorsal scale; large disc in lower eyelid; vertically elongate ear opening with 2 round anterior lobules; black mark centred on ear opening. Breeding males uniform pale brown with reddish orange paravertebral, upper lateral and lower lateral stripes. Sides of head and neck dark greyish brown with white spots. Females, juv., non-breeding males brown with broad, dark brown dorsal zone enclosing pale dashes, each dark-edged anteriorly, between obscure pale vertebral and dorsolateral stripes. Midlateral stripe greyish brown. Pale line curves from nostril, under eye to ear.

Endemic. Heaths and low woodlands on white sands between mouth of McIvor R. and Bathurst Head, inland to Jowalbinna Stn, and on Eagle and Lizard Is, CYP.

Carlia inconnexa. Female. Whitsunday Is.

Carlia inconnexa. Breeding male. Whitsunday Is.

Whitsunday Rainbow Skink

Carlia inconnexa

SVL 52mm

Mostly 2 (sometimes mixed with 3) keels on hexagonal dorsal scales; large transparent disc occupying more than half lower eyelid; usually vertically elliptic ear-opening with 1–2 round lobules on anterior margin; usually 5 supraciliary scales. Grey to brown with paravertebral stripes of black blotches and white flecks. Breeding males have black heads and blue throats with prominently black-edged scales.

Endemic. CQC, on Hayman, Hook, Whitsunday and Lindeman Islands. Recorded from stony soils.

Carlia jarnoldae. Breeding male. Bowling Green Bay NP.

Carlia jarnoldae. Female. Bowling Green Bay NP.

Lined Rainbow Skink

Carlia jarnoldae

SVL 49mm

Three moderately strong keels on each hexagonal dorsal scale; large disc in lower eyelid; horizontally elongate ear opening with small pointed anterior lobule and smaller lobules on remaining margin. Breeding males brown above with 5–7 narrow, dark dorsal lines, black upper flanks enclosing numerous small, sharp, blue spots, bright orange mid to lower flanks and forelimbs, and pale green lips, sides of neck and ventral surfaces. Females, juv., non-breeding males brown above with black upper flanks and broad white midlateral stripe from upper lips through ear opening and along body.

Endemic. Dry sclerophyll forests and woodlands, usually on well-drained stony soils and outcrops in nthn BB, EIU, WT, CYP.

Closed-litter Rainbow Skink
Carlia longipes
SVL 68mm

Smooth or with 3 very weak keels on dorsal scales each with curved hind edges; small disc in lower eyelid; ear-opening usually round, occasionally vertical or rarely horizontal, completely surrounded by sharply pointed lobules; large size; robust build.

Breeding males are brown with narrow pale dorsolateral stripe from above eye to behind forelimb, black midlateral zone from head to shoulder, sharply contrasting with pale lower neck and throat, and rich orange flanks. Females and juv. have narrow pale stripe from under eye to above ear, pale dorsolateral and midlateral stripes and dark upper lateral zone, fading posteriorly.

Endemic. WT and se. CYP, along coast between Cape Melville and Gordonvale.

Carlia longipes. Female. Cooktown.

Carlia longipes. Breeding male. Cooktown.

Striped Rainbow Skink
Carlia munda
SVL 44mm

Virtually smooth dorsal scales with curved hind edges; large disc in lower eyelid; horizontally elongate ear opening with few small lobules on upper edge, occasionally smaller ones on other margins; disrupted white midlateral stripe extending from upper lip to top of ear opening, then continuing back from bottom of ear opening. Brown to grey with dark and pale spots and flecks. On breeding males, flanks flushed with black from head to about forelimb, with reddish orange from forelimb to hindlimb, scales on sides of neck and throat prominently edged with black.

Carlia munda. Breeding male. Shoalwater Bay.

Dry sclerophyll forests and woodlands, mainly on heavy to stony soils, over all but most arid parts of Qld. Also NT, WA.

Open-litter Rainbow Skink
Carlia pectoralis
SVL 47mm

Three strong keels on each hexagonal dorsal scale; large transparent disc in lower eyelid; ear-

Carlia pectoralis. Female. Carnarvon Range.

Carlia quinquecarinata. Darnley Island, Qld. D. Stanton

Carlia pectoralis. Breeding male. Woodgate area.

opening usually round, with rounded anterior lobule and sharp to blunt lobules around margin; 5 supraciliary scales. Brown with paravertebral rows of black and white dots. Breeding male has orange upper and sometimes lower lateral stripes from forelimb to groin, orange flush on chest, and blue throat, neck and chin with strongly black-edged scales. Females and juv. have narrow pale midlateral stripe with diffuse dark edges, prominent anteriorly but breaking into flecks at anterior or mid flanks.

Endemic. Dry open forests, often with grass tussocks in SEQ, BB and CQC, from Brisbane inland to Carnarvon Ra. and n. to Blackdown Tableland and Shoalwater Bay.

Five-carinated Rainbow Skink

Carlia quinquecarinata

SVL 65mm

Smooth to weakly carinated dorsal scales; vertically elongate ear-opening with sharp lobules on all margins, small and deeply recessed on posterior margin; robust build. Brown with large, longitudinally-aligned dark-edged pale dashes, and sometimes a broken pale midlateral line.

Darnley Is. and probably Murray Is, ne. Torres Strait. Also sthn NG. Recorded in semi-deciduous notophyll vine forest on basalt.

Carlia rhomboidalis. Magnetic Island.

Blue-throated Rainbow Skink

Carlia rhomboidalis

SVL 57mm

Smooth dorsal scales with curved hind edges; interparietal scale fused to frontoparietal scale; round to vertically elongate ear-opening with 1–3 large pointed anterior lobules, and sometimes smaller sharp ones around edges; blue and red throat. Brown with dark flecks along paravertebral zone, gold to pale brown dorsolateral stripe, dark flanks usually enclosing wavy to broken pale midlateral stripe. Top of head rich copper. During breeding season (summer), chin and lips bright blue, throat prominently flushed with red.

Endemic. Rainforests, wet sclerophyll forests and margins in CQC, nthn BB, n. to Magnetic Is.

Carlia rostralis. Male. Davies Creek.

Carlia rostralis. Female. Isabella Falls.

Black-throated Rainbow Skink

Carlia rostralis

SVL 70mm

Smooth or with 3 weak carinations on dorsal scales, each with curved hind edges; small disc in lower eyelid; vertically elongate ear opening with large pointed anterior lobules, sometimes smaller lobules on other margins; large size, robust build. Females and juv. brown above with scattered dark flecks and prominent pale dorsolateral stripes. Anterior flanks, from snout to behind forelimb, black with pale midlateral stripe. Posterior flanks brown, sometimes suffused with red. Breeding males much brighter; very prominent cream dorsolateral stripe from above eye to behind forelimb, black upper flanks usually fading to dark brown at mid or posterior body, very prominent pale midlateral stripe from lips to about forelimb, bright reddish orange lower flanks, boldest anteriorly, black chin and throat.

Endemic. Dry forests, woodlands, vine thickets, outcrops in nthn BB, WT, EIU; isolated records from Kowanyama in wstn CYP.

Carlia rubigo. Female. Magnetic Is.

Carlia rubigo. Breeding male. Hallett's State Forest.

Orange-flanked Rainbow Skink

Carlia rubigo

SVL 44mm

Three strong keels on each hexagonal dorsal scale; large transparent disc in lower eyelid; ear-opening vertically elliptical, with 1–2 rounded anterior lobules and smaller rounded lobules on other margins; 5 supraciliary scales. Brown with paravertebral rows of black and white dots on posterior body. Breeding male has orange to copper flush on flanks sometimes reaching back and forelimbs, and blue with dark speckling on throat, jaw and sides of neck. Female and juv. has white midlateral stripe from nostril to forelimb or mid flanks, breaking into dashes posteriorly.

Endemic. BB, CQC, EIU and sthn WT. Dry forests and rocky areas, in estn interior s. to Texas and St George areas, reaching coast and islands between about Mackay and Townsville and extending n. to w. of Cairns.

Carlia rubrigularis. Josephine Falls.

Southern Wet Tropics Red-throated Rainbow Skink

Carlia rubrigularis

SVL 55mm

Smooth dorsal scales with curved hind edges; interparietal scale fused to frontoparietal scale; round ear-opening with 1–2 large pointed anterior lobules and smaller pointed lobules on remaining edges; red throat. Rich brown with black flecks mainly on paravertebral zone, a narrow pale dorsolateral stripe, dark upper flanks, an indication of pale midlateral stripe and copper on top of head. Pink on chin and throat, brightest on breeding male. Distinguishable from *C. crypta* only by distribution and genetics.

Endemic. Rainforests and margins in sthn Wet Tropics, from sthn end of Paluma Ra., n. to a line appr. from Mareeba, through Lake Tinaroo, crest of Lamb Ra. to Cairns.

Carlia schmeltzii. Breeding male. Isla Gorge.

Schmeltz's Rainbow Skink

Carlia schmeltzii

SVL 69mm

Two or 3 strong keels (mainly 3 s. of Townsville, 2 n. and w., and mix of both in Townsville area) on hexagonal dorsal scales; small transparent disc in lower eyelid; vertically elongate ear opening, usually with 2 large, squarish anterior lobules. Brown above with broad pale dorsolateral stripe, grey on flanks with broadly black-edged scales. Breeding males have pale, conspicuously black-edged scales on sides of head and neck, bright reddish orange lateral flush from forelimb to mid or posterior body. Pops of small, patternless lizards with 2 keels on Cape York once considered separate sp., *C. prava.*

Endemic. Dry woodlands, rocky ranges and scree slopes in SEQ, BB, CQC, DEU, EIU, WT, far nthn GUP, CYP n. to Coen.

Robust Rainbow Skink

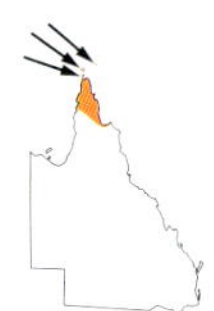

Carlia sexdentata

SVL 63mm

Smooth or with 3 very weak keels on dorsal scales, each with curved hind edges; small disc in lower eyelid; ear-opening usually vertical, occasionally round, with well-developed

Carlia sexdentata. Female. Moa Is.

Carlia sexdentata. Breeding male. Iron Range.

sharp lobules on anterior margin and moderately to poorly developed lobules on surrounding margins; large size; robust build. Breeding male has weak pale dorsolateral stripe from above eye to behind forelimb, dark grey upper lateral zone from head to shoulder, merging with rich orange flanks. Female and juv. has narrow pale stripe from under eye to above ear, pale dorsolateral and midlateral stripes, dark upper flanks fading posteriorly, usually broken by obscure pale vertical bars in front of forelimb and enclosing irregular pale dashes along body

Variety of habitats incl. vine thickets and disturbed areas in CYP and Torres Strait is. s. to Princess Charlotte Bay. Also NT.

Carlia storri. Cooktown.

Storr's Rainbow Skink

Carlia storri

SVL 46mm

Two strong keels on hexagonal dorsal scales; small transparent disc in lower eyelid; round ear opening margined by pointed lobules. Pale brown with paler vertebral stripe to base of tail, dorsolateral stripes onto tail, midlateral stripes from near forelimb to just before hindlimb. On breeding males stripes replaced with orange flush on limbs and tail.

Dry sclerophyll forests, woodlands in WT, CYP, through Torres Strait is. to sthn NG.

Southern Rainbow Skink

Carlia tetradactyla

SVL 64mm

Smooth to very weakly striated dorsal scales with curved hind edges; small disc in lower eyelid; round ear opening with large blunt

Carlia tetradactyla. Breeding male. Oakey.

anterior lobule; large size, robust build. Pale greyish brown with broad blackish brown dorsal zone enclosing pale spots or dashes and broad pale dorsolateral stripe. Breeding males develop bright reddish orange upper and lower lateral stripes, pale bluish green wash over flanks and ventral surfaces.

Dry sclerophyll forests, grassy woodlands, grasslands of NET, sthn BB, between about Stanthorpe and Oakey areas. Also estn NSW to nthn Vic. Only member of essentially dry-tropical genus adapted to temperate climates, being restricted to coolest sthn parts of State.

Carlia vivax. Female. Coominya.

Lively Rainbow Skink

Carlia vivax

SVL 47mm

Two strong keels on each hexagonal dorsal scale; very large disc in lower eyelid; vertically elongate ear opening with large blunt anterior lobule. Brown to greyish brown with 2 broad dark dorsal zones enclosing pale dashes, pale dorsolateral stripe from neck onto tail, sharp-edged white midlateral stripe from upper lips through ear to hindlimb. Breeding males have broad, prominent orange lateral stripes, pale blue chin and throat with dark speckling.

Variety of dry wooded habitats in SEQ, NET, BB, CQC, EIU, WT, CYP, n. to Prince of Wales Is. Also estn NSW.

Carlia vivax. Breeding male. Kurwongbah.

C. Hoskin

Carlia wundalthini. Male. Melville Range.

Cape Melville Rainbow Skink

Carlia wundalthini

SVL 49mm

Smooth dorsal scales with curved hind edges; interparietal scale fused to frontoparietal scale; round to horizontally elongate ear-

C. Hoskin

Carlia wundalthini. Female. Melville Range.

opening with large sharp anterior lobule and similar lobules around top or sometimes whole margin; breeding colours restricted to orange flush on sides of neck and flanks. Brown, sometimes with 2 series of dark dorsal flecks and copper head. Breeding males have orange wash on sides of neck and flanks but throat unmarked or with only faint indication. Females have broken midlateral line of pale dots.

Endemic. Isolated upland rainforest atop the boulder fields of Melville Range, CYP.

Genus *COERANOSCINCUS*

2 Aust. spp.; both occur in Qld. Limbs either absent or greatly reduced with 3 digits; ear opening represented by scaly depression; small eye with movable lower eyelid lacking transparent disc; round snout (not pointed or wedge-shaped); smooth glossy scales; large size; strikingly patterned, banded or striped juv. but almost patternless adults.

Rainforest inhabitants in widely separated blocks: WT and SEQ. Secretive burrowers, dwelling in thick moist leaf litter, soil and wood mulch. Mainly nocturnal but sometimes encountered active by day during overcast and wet weather. It is believed their pointed, curved teeth may be an adaptation to grasp earthworms. At least one sp. an egglayer.

Coeranoscincus frontalis. Adult. Walter Hill Ra.

Coeranoscincus frontalis. Juvenile. Mt Bellenden Ker.

Limbless Snake-tooth Skink

Coeranoscincus frontalis

SVL 290mm

No limbs. Juv. has white head with black patches on eyes and ear depressions, dark greyish brown back with narrow white dorsal lines, broad white upper lateral stripe continuous with white head, black lower lateral stripe. Adult sombre brown, only a trace of bold juv. pattern.

Endemic. Tropical rainforests of WT, between Thornton Peak and Paluma Ra.

Three-toed Snake-tooth Skink

Coeranoscincus reticulatus

SVL 250mm

Four very short limbs, each with 3 digits. Juv. cream to brown, palest on head, dark patch on crown, dark patches on eyes and ear depressions, conspicuous dark bands across neck and forebody, becoming weaker posteriorly. Adults brown to orange with indications of dark eye and ear markings, pale snout, sometimes remnants of bands.

Subtropical rainforests in McPherson, Main and Connondale Ra., n. to Fraser Is., SEQ; absent from apparently suitable habitats in D'Aguilar Ra. Also ne. NSW. Mainly elevated habitats, though Fraser Is. and adj. Cooloola region are lowland sites. Clutches of 2–6 eggs.

Coeranoscincus reticulatus. Adult and juvenile. Cunningham's Gap.

Genus *COGGERIA* Monotypic genus.

Fraser Island Sand Skink

Coggeria naufragus

SVL 127mm

Long-bodied with very short limbs, each with 3 digits; movable lower eyelid with no transparent disc; ear opening represented by scaly depression; pointed wedge-shaped snout; smooth glossy scales; large size. Very pale brown above with obscure, narrow, dark longitudinal lines, and prominent dorsolateral line of black spots clearly delineating pale back from grey, black-flecked flanks and ventral surfaces.

Endemic. Restricted to Fraser Is., SEQ. Secretive burrower. Diverse forest and heath habitats on sandy soils.

Coggeria naufragus. Central Stn, Fraser Is.

Rarely encountered under debris; most specimens collected in pit-traps, suggesting it may dwell deep in substrate but forage regularly on the surface.

Genus *CONCINNIA*

7 Aust spp.; all occur in Qld. Well-developed limbs overlap when pressed to side of body, all with 5 digits; movable lower eyelid lacking a transparent disc; no lobules on ear-opening; parietal scales in contact; frontoparietals divided; all distal subdigital lamellae undivided; smooth, shiny scales.

Moist estn areas. Also estn NSW. Most occupy cavities in tree trunks, logs and rock crevices, sometimes residing in rockeries and other garden habitats. All are diurnal livebearers. Diets consist mainly of arthropods.

KEY TO *CONCINNIA*

1	Supranasal scales present (**a**)	***ampla***
	No supranasal scales	2
2(1)	Prefrontal scales in broad contact (**b**)	***tigrina***
	Prefrontal scales in point contact (**c**) or separated	3
3(2)	Upper secondary temporal scale overlaps lower (**d**)	4
	Lower secondary temporal scale overlaps upper (**e**)	6
4(3)	Subdigital lamellae 22–27; often a dark vertebral streak on nape	***tenuis***
	Subdigital lamellae 17–23; no dark vertebral streak on nape.	5
5(4)	Midbody scales in 28–32 rows; coppery brown with dark brown to black dorsal flecks	***brachysoma***
	Midbody scales in 32–38 rows; pale brown with darker brown dorsal flecks	***sokosoma***
6(3)	Paravertebral scales 64 or fewer; se. to mid-e.Qld	***martini***
	Paravertebral scales more than 64; WT, above 1000 m	***frerei***

Concinnia ampla. Finch Hatton.

Concinnia brachysoma. Mt Abbott.

Lemon-barred Forest-skink
Concinnia ampla
SVL 115mm

Pair of supranasal scales; one lower labial scale contacts postmental scale; banded pattern; robust build. Dark olive brown, irregular narrow bands of paler scales on body and tail, prominent dark patch above forelimb, bright lemon yellow ventral surfaces.

Endemic. Confined to rainforests of CQC, in vicinity of Eungella NP, Finch Hatton, Mt Blackwood and Conway SF. Often seen basking on streamside boulders or amongst root systems of larger trees. Unusual amongst skinks in frequently sleeping on exposed sites such as rock faces.

Northern Bar-sided Skink
Concinnia brachysoma
SVL 74mm

Narrow bands on tail; 2 lower labial scales contact postmental scale; 28–32 midbody scale rows; upper secondary temporal scale overlaps lower; 17–23 lamellae under 4th toe. Coppery brown above with numerous small dark blotches tending to concentrate on either side of midline, broad black lateral stripe with deeply zigzagging edges, sometimes broken into vertical bars.

Endemic. From about Coen in CYP, through WT, EIU, BB, CQC to about Gayndah, SEQ. Mainly inhabits rocky areas, often associated with moist, well-vegetated pockets within drier terrain. Shelters in rock crevices, in burrows under rocks, tree hollows.

Concinnia frerei. Mt Bartle Frere.

Bartle Frere Bar-sided Skink

Concinnia frerei

SVL 66mm

Narrow bands on sides of tail; 2 lower labial scales contact postmental scale; lower secondary temporal scale overlaps upper. Dark brown above with black, deeply notched upper lateral zone continuous with series of irregular, finger-like, broken dorsal bands. Ventral surfaces pale green.

Endemic. Elevations above 1400 m on Mt Bartle Frere and adjacent Mt Bellender Ker, and 110 km ne. at 1200 m elevation on Mt Lewis. Recorded from exposed granite boulders surrounded by dense rainforest, and in cavities on mossy tree trunks. Habitat often cloaked in mist, subject to year-round low temperatures.

Concinnia martini. Verrierdale.

Martin's Skink

Concinnia martini

SVL 70mm

Narrow bands on tail; 2 lower labial scales contact postmental scale; lower secondary temporal scale overlaps upper. Coppery brown with mosaic of irregular black flecks over back and broad, deeply notched, black upper lateral stripe.

Variety of rocky, forested and disturbed habitats in NET, SEQ, BB, CQC. Also ne. NSW. Shelters in rock crevices, cavities in timber. Often occurs near similar *C. tenuis*, tending to be less arboreal (preferring logs and rocks to standing trees) within zone of overlap. Frequently occurs in dwellings, in cool rooms such as laundries, and in rockeries.

Concinnia sokosoma. Hallett's State Forest.

Stout Bar-sided Skink

Concinnia sokosoma

SVL 79mm

Narrow bands on tail; 2 lower labial scales contact postmental scale; 32–38 midbody scale rows; upper secondary temporal scale overlaps lower; 19–23 lamellae under 4th toe. Pale brown with dark dorsal blotches, either scattered or aligned along paravertebral region, deeply notched to broken upper lateral stripe.

Endemic. Mainly rock-inhabiting, in crevices associated with sheltered moist areas in drier outcrops and ra. of sthn WT, EIU, far nthn DEU, CQC, BB, s. to Injune area.

Concinnia tenuis. Cooloola NP.

Barred-sided Skink

Concinnia tenuis

SVL 85mm

Narrow bands on tail; usually narrow dark line on nape; 2 lower labial scales contact postmental scale; upper secondary temporal scale overlaps lower; 22–27 lamellae under 4th toe. Coppery brown to pale brown above, with small black blotches and dark upper lateral stripe usually broken into series of vertical bars. Dark markings usually more reduced (sparser on back, more broken on flanks) than on similar *C. martini* and *C. brachysoma.*

Variety of forested, rocky and disturbed habitats incl. gardens and houses in NET, BB, SEQ, CQC, WT. Also estn NSW. Often occupies cavities in standing timber and is seen with head protruding; shelters in rock crevices and gaps in stone walls.

Concinnia tigrina. Malanda Falls.

Yellow-blotched Forest-skink

Concinnia tigrina

SVL 85mm

Dorsolateral row of pale blotches; 2 lower labial scales contact postmental scale; broad contact between prefrontal scales. Shades of brown, with dark dorsal bands, when present, narrow on back and broadening to form dark dorsolateral blotches which alternate with distinctive dorsolateral series of cream to orange blotches. Ventral surfaces yellow to pale green.

Endemic. Mainly mid to high-altitude rainforests of WT, from Shipton's Flat s. to Cardwell Ra. Occupies rotting logs; often encountered beside popular walking tracks resting with head protruding from cavity. In high-altitude heaths it often uses rock crevices.

Snake-eyed Skinks, Wall Lizards, Fence Skinks Genus *CRYPTOBLEPHARUS*

At least 55 spp. from Africa to Pacific; 23 in Aust.; 9 occur in Qld. Flat-bodies with long limbs, each with 5 long, slender digits; lower eyelid fused to form a large fixed spectacle surrounded by granular scales, usually the upper 3 greatly enlarged; parietal scales in contact; interparietal and frontoparietal scales fused to form large, kite-shaped shield; frontal and prefrontal scales about equal in size.

Statewide, apart from some high-altitude and rainforest areas. All mainland States. These extremely swift, sun-loving lizards move with ease over vertical rock faces, tree trunks and walls. They favour dry habitats and tolerate desiccation, a likely factor in their successful dispersal across far-flung oceanic islets. Egglayers.

KEY TO *CRYPTOBLEPHARUS*

1	Supraciliary scales **(a)** usually 6	2
	Supraciliary scales usually 5	3
2(1)	Plantar scales rounded (cobblestone-like) **(b)**	***metallicus***
	Plantar scales pointed **(c)**	***australis***
3(1)	Plantar scales rounded (cobblestone-like) **(b)**	4
	Plantar scales pointed **(c)**	***pannosus***
4(3)	Usually 18 scales along dorsal surface of 4th toe; laterodorsal stripes comprise longitudinal rows of sharp white dashes; Cape Melville on CYP only	***fuhni***
	Usually 16 or fewer scales along dorsal surface of 4th toe; laterodorsal stripes (if present) continuous	5
5(4)	Midbody scales usually 26; 4th toe subdigital lamellae usually 20; coastal rocks only	***litoralis***
	Midbody scales usually 24 or fewer; 4th toe subdigital lamellae usually 19 or fewer	6
6(5)	Reddish; no pale laterodorsal stripes	***zoticus***
	Brown, grey or black, with pale laterodorsal stripes	7
7(6)	Midbody scales usually 22; paravertebral scales usually 47	***virgatus***
	Midbody scales usually 24; paravertebral scales usually 50	8
8(7)	Plantar scales dark; pale laterodorsal stripes narrow (half to three quarters laterodorsal scale width); smooth-edged	***pulcher***
	Plantar scales pale; pale laterodorsal stripes broad (about laterodorsal scale width); ragged-edged	***adamsi***

Cryptoblepharus adamsi. Bowen River via Collinsville.

Adam's Snake-eyed Skink

Cryptoblepharus adamsi

SVL 37mm

Usually 5 supraciliary scales; rounded (cobblestone-like) plantar scales; prominent pale laterodorsal stripes. Narrow dark brown to blackish vertebral zone and prominent broad ragged-edged pale dorsal stripes with broad, black, ragged inner margins.

Endemic. CQC, WT and EIU, from Mt Molloy s. to Mt Larcom and inland to w. of Mt Surprise. Arboreal in open forests, woodlands, parks and gardens.

Cryptoblepharus australis. Taroom area.

Inland Snake-eyed Skink

Cryptoblepharus australis

SVL 46mm

Usually 6 supraciliary scales; acute, pointed plantar scales; ragged longitudinally-aligned pattern. Greyish with ragged pattern incl. broad dark vertebral zone, broad pale laterodorsal zones with dark inner edges comprising dark flecks aligned to form 2 narrow, broken, ragged paravertebral stripes.

Most of Qld interior, n. to about Barkly Hwy. Also interior of all mainland states. Mainly arboreal, frequently occupying human structures.

Cryptoblepharus fuhni. Cape Melville.

Fuhn's Snake-eyed Skink

Cryptoblepharus fuhni

SVL 47mm

Usually 5 supraciliary scales; round dark plantar scales; very flat head and body and extremely long limbs. Black to very dark brown with prominent white dots and dashes forming broken dorsolateral lines and parallel lateral rows.

Endemic. Rock-inhabiting, restricted to piles of black boulders of Melville Ra., estn CYP.

Coastal Snake-eyed Skink

Cryptoblepharus litoralis

SVL 51mm

Usually 5 supraciliary scales; rounded, dark plantar scales; longitudinally-aligned pattern, and long limbs. Pale to dark brown with dark and pale flecking, ragged-edged pale

Cryptoblepharus litoralis. Prince of Wales Is.

dorsolateral stripes with irregular broken inner edges.

Exposed rocky coastline from Gladstone district, n. through Torres Strait is. to sthn NG. Also nthn NT. Seldom found more than 100m inland. Forages for invertebrates so close to water that it must sometimes dodge incoming waves.

Metallic Snake-eyed Skink

Cryptoblepharus metallicus

SVL 47mm

Usually 6 supraciliary scales; rounded, usually callused plantar scales; ragged longitudinally-aligned pattern. Grey to greyish brown with ragged, prominent to obscure pattern incl. broad dark vertebral zone about as wide as paravertebral scales, and pale laterodorsal zones with dark inner edges comprising dark flecks aligned to form 2 narrow broken paravertebral stripes.

Nthn and estn Qld. Also nthn NT and WA. Largely arboreal, but frequently utilises human structures and sometimes rocks.

Cryptoblepharus metallicus. Bowen River via Collinsville.

Cryptoblepharus metallicus. Prince of Wales Island.

Cryptoblepharus pannosus. Glenmorgan.

Ragged Snake-eyed Skink

Cryptoblepharus pannosus

SVL 41mm

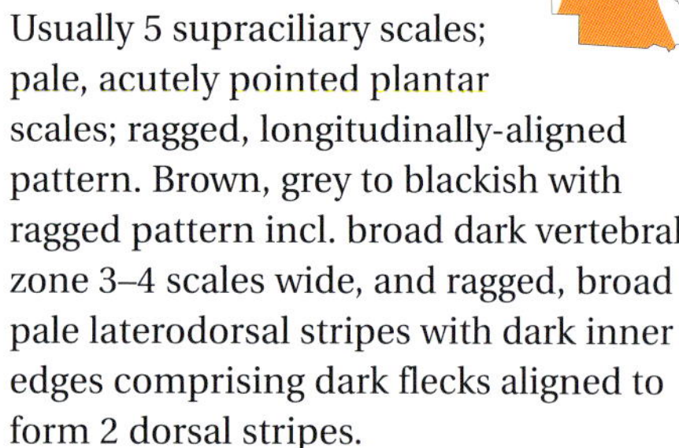

Usually 5 supraciliary scales; pale, acutely pointed plantar scales; ragged, longitudinally-aligned pattern. Brown, grey to blackish with ragged pattern incl. broad dark vertebral zone 3–4 scales wide, and ragged, broad pale laterodorsal stripes with dark inner edges comprising dark flecks aligned to form 2 dorsal stripes.

Most of Qld with possible exception of SEQ, CQC and far wstn arid zones. Also from NSW to estn SA and far nthn Vic. Mainly arboreal but extremely versatile, occurring on rocks and frequently occupying human structures.

Cryptoblepharus pulcher. Kurwongbah.

Elegant Snake-eyed Skink

Cryptoblepharus pulcher

SVL 41mm

Usually 5 supraciliary scales; round, dark plantar scales; usually 24 midbody scales; usually 50 paravertebral scales; smooth, longitudinally-aligned pattern. Brown to brownish-black with narrow dark vertebral zone about 1 scale wide, and prominent, narrow, smooth-edged pale laterodorsal stripes, their inner edges comprising broad black dorsal stripes.

Variety of tree, rock and urban habitats of SEQ, NET, BB, CQC, EIU and WT, n. to Ingham area. Also estn NSW. Another ssp. in sthn SA and sthn WA. Arboreal and rock-inhabiting, often extending onto human structures.

Striped Snake-eyed Skink

Cryptoblepharus virgatus

SVL 39mm

Usually 5 supraciliary scales; round, pale plantar scales; usually 22 midbody scales; usually

Cryptoblepharus virgatus. Cooktown.

47 paravertebral scales; smooth, longitudinally-aligned pattern. Grey to greyish-brown with broad dark vertebral zone about 2 scales wide, and prominent, narrow, smooth-edged pale laterodorsal stripes, their inner edges comprising prominent black dorsal stripes.

Endemic. Nthn BB, WT and CYP, from about Townsville to Torres Strait. Arboreal but often inhabiting humans structures.

Cryptoblepharus zoticus. Lawn Hill NP.

Agile Snake-eyed Skink

Cryptoblepharus zoticus

SVL 38mm

Usually 5 supraciliary scales; rounded plantar scales; flat head and body; speckled or blotched pattern. Reddish with random, weak to prominent irregular dark flecks and spots.

NWH, se. to Mary Kathleen area. Also adj. ne. NT. Exclusively rock-inhabiting on escarpments, hills and outcrops.

Striped Skinks
Genus *CTENOTUS*

106 Aust. spp.; 45 occur in Qld. Well-developed limbs, each with 5 digits; long slender tails; movable lower eyelid with no transparent disc; parietal scales in contact; frontoparietals divided; smooth body scales; obvious ear openings with anterior lobules present on all but *C. brevipes*. Most are patterned with prominent longitudinal stripes, often with pale lateral spots.

Throughout Qld, incl. most towns and cities but not closed rainforests and wetlands. Extremely swift, diurnal lizards of dry open terrain, attaining greatest diversity in arid zones, particularly spinifex deserts, venturing onto exposed, open ground much more readily than most skinks. Egglayers.

KEY TO *CTENOTUS*

1 3 supraocular scales ..... ***spaldingi***
4 supraocular scales 2

2(1) Pattern comprises dark-edged pale ocelli ..... ***pantherinus***
Pattern variable but never solely comprising dark-edged pale ocelli 3

3(2) Scales along top of 4th toe have transverse sutures along almost entire digit (**a**) 4
Scales along top of 4th toe have oblique sutures on at least basal quarter to third of digit (**b**) 31

4(3) Pattern consists only of stripes (either few or numerous), with spots absent, or sometimes present on juveniles 5
Pattern plain, spotted, or featuring both stripes and spots 15

5(4) 18 or more pale stripes 6
Much fewer than 18 pale stripes 7

6(5) Lateral stripes, including pale midlateral stripe, extend forward beyond forelimb ..... ***dux***
All lateral stripes fragmented in front of forelimb ..... ***ariadnae***

7(5) Black vertebral stripe from nape at least to about hips; ear-lobules present 8
Black vertebral stripe absent or reduced to a dash on nape or foreback; ear-lobules absent or barely discernible ..... ***brevipes***

8(7) Pale upper lateral stripe or row of dashes between dorsolateral and midlateral stripes ..... ***lateralis***
No pale upper lateral stripe 9

9(8) Little or no pale edge to dark vertebral stripe ..... ***brachyonyx***
Sharp pale paravertebral stripes edging dark vertebral stripe 10

10(9) Subdigital lamellae narrowly to broadly callose (**c**) 11
Subdigital lamellae finely keeled and mucronate (**d** and **e**) 12

11(10) Pale stripes on body 10–12 (rarely 8); rocky substrates ..... ***decaneurus***
Pale stripes on body 6 (occasionally 8); sandy substrates ..... ***piankai***

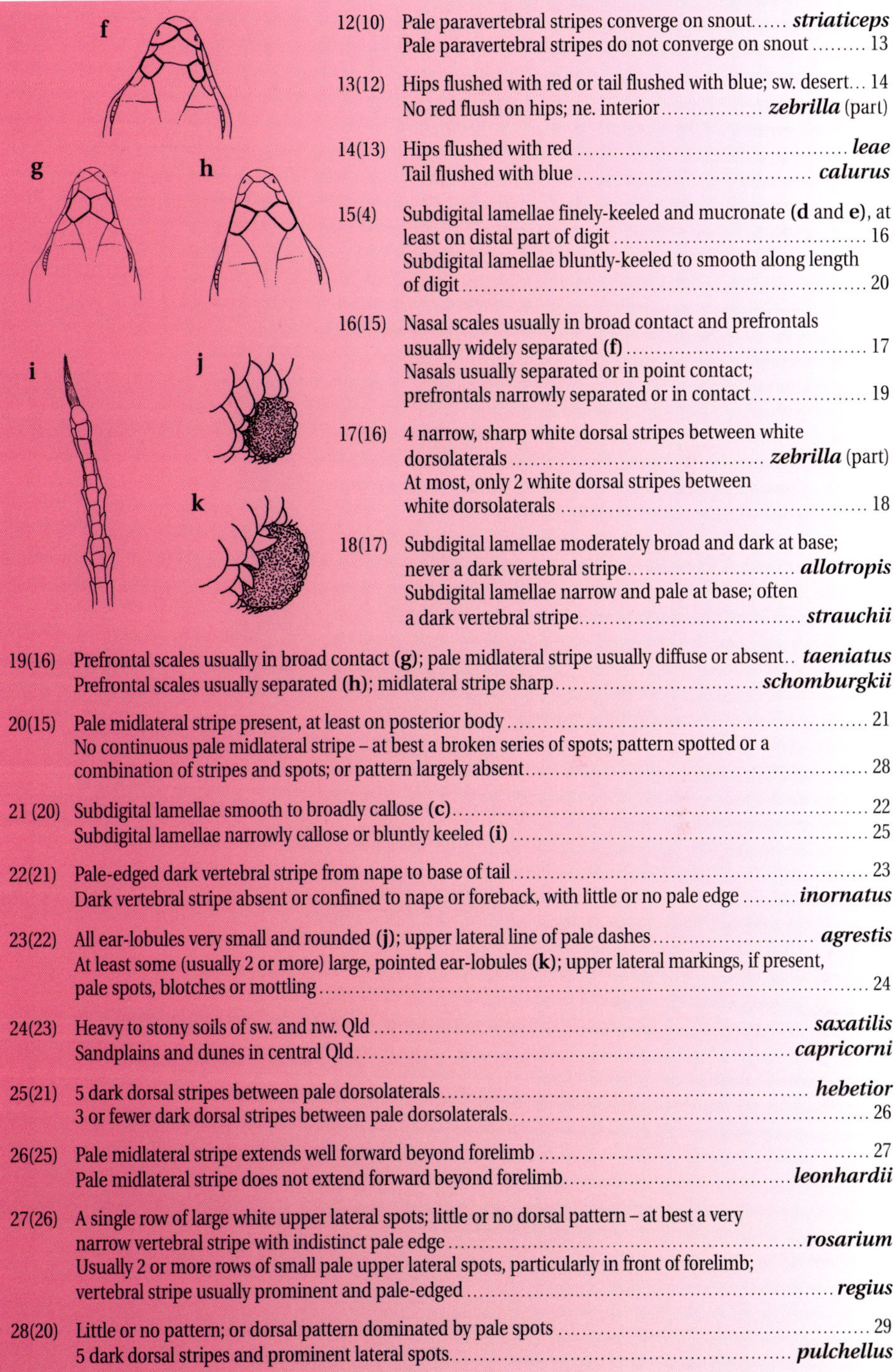

12(10) Pale paravertebral stripes converge on snout...... ***striaticeps***
Pale paravertebral stripes do not converge on snout......... 13

13(12) Hips flushed with red or tail flushed with blue; sw. desert... 14
No red flush on hips; ne. interior................. ***zebrilla*** (part)

14(13) Hips flushed with red .... ***leae***
Tail flushed with blue .... ***calurus***

15(4) Subdigital lamellae finely-keeled and mucronate (**d** and **e**), at least on distal part of digit 16
Subdigital lamellae bluntly-keeled to smooth along length of digit 20

16(15) Nasal scales usually in broad contact and prefrontals usually widely separated (**f**) 17
Nasals usually separated or in point contact; prefrontals narrowly separated or in contact................... 19

17(16) 4 narrow, sharp white dorsal stripes between white dorsolaterals .... ***zebrilla*** (part)
At most, only 2 white dorsal stripes between white dorsolaterals 18

18(17) Subdigital lamellae moderately broad and dark at base; never a dark vertebral stripe.......................... ***allotropis***
Subdigital lamellae narrow and pale at base; often a dark vertebral stripe.................................... ***strauchii***

19(16) Prefrontal scales usually in broad contact (**g**); pale midlateral stripe usually diffuse or absent.. ***taeniatus***
Prefrontal scales usually separated (**h**); midlateral stripe sharp................................ ***schomburgkii***

20(15) Pale midlateral stripe present, at least on posterior body 21
No continuous pale midlateral stripe – at best a broken series of spots; pattern spotted or a combination of stripes and spots; or pattern largely absent.. 28

21 (20) Subdigital lamellae smooth to broadly callose (**c**)... 22
Subdigital lamellae narrowly callose or bluntly keeled (**i**) 25

22(21) Pale-edged dark vertebral stripe from nape to base of tail 23
Dark vertebral stripe absent or confined to nape or foreback, with little or no pale edge .... ***inornatus***

23(22) All ear-lobules very small and rounded (**j**); upper lateral line of pale dashes.......................... ***agrestis***
At least some (usually 2 or more) large, pointed ear-lobules (**k**); upper lateral markings, if present, pale spots, blotches or mottling.. 24

24(23) Heavy to stony soils of sw. and nw. Qld .... ***saxatilis***
Sandplains and dunes in central Qld.. ***capricorni***

25(21) 5 dark dorsal stripes between pale dorsolaterals... ***hebetior***
3 or fewer dark dorsal stripes between pale dorsolaterals... 26

26(25) Pale midlateral stripe extends well forward beyond forelimb 27
Pale midlateral stripe does not extend forward beyond forelimb.................................... ***leonhardii***

27(26) A single row of large white upper lateral spots; little or no dorsal pattern – at best a very narrow vertebral stripe with indistinct pale edge .... ***rosarium***
Usually 2 or more rows of small pale upper lateral spots, particularly in front of forelimb; vertebral stripe usually prominent and pale-edged .... ***regius***

28(20) Little or no pattern; or dorsal pattern dominated by pale spots 29
5 dark dorsal stripes and prominent lateral spots.. ***pulchellus***

29(28) Little or no pattern; sand dunes and red loamy flats in far w. Qld ..... ***helenae***
Dorsal pattern dominated by pale spots; clay or gravelly soils in sw. and central Qld 30

30(29) Nasal scales usually in contact; 40 or more midbody scale rows ..... ***schevillii***
Nasal scales usually separated; fewer than 40 midbody scale rows ..... ***astarte***

31(3) Scales along top of 4th toe have oblique sutures along almost entire digit; CYP only 32
Scales along top of 4th toe have oblique sutures along at least basal quarter to two thirds of digit, and with transverse sutures on distal portion (rarely oblique sutures along entire digit in SEQ) 33

32(31) Back plain ..... ***quinkan***
Back striped ..... ***rawlinsoni***

33(31) Pattern either striped or plain, but no pale spots, though upper lateral stripe may break into dashes 34
Pattern either striped or fragmented, but pale lateral spots present 40

34(33) Vertebral stripe absent, at best a narrow line on nape and foreback 35
Vertebral stripe present 36

35(34) Body patternless; sandy coastal and offshore SEQ ..... ***robustus*** (part)
Black upper flanks and sharp white upper lateral stripe ..... ***terrareginae***

36(34) Pale upper lateral stripe between midlateral and dorsolateral stripes ..... ***eutaenius***
No pale stripe within upper flanks 37

37(36) Upper flanks usually black 38
Upper flanks grey to brown; MGD only ..... ***joanae*** (part)

38(37) At most, 4 pale dorsal stripes – dorsolaterals and sometimes paravertebrals ..... ***ingrami***
6 pale stripes, with dorsal colour contracted to form an additional stripe 39

39(38) Snout pointed in profile; pale midlateral stripe continuous over or around ear-opening; flanks unspotted ..... ***taeniolatus***
Snout round in profile; pale midlateral stripe often discontinuous at anterior edge of ear-opening; white lateral spots usually present ..... ***eurydice*** (part)

40(33) Continuous pale midlateral stripe present, at least on posterior body, or if absent, stripes along sides of tail sharp and black 41
Pale midlateral stripe absent – at best a row of spots form an irregular stripe on posterior body; tail stripe diffuse, stippled or absent 48

41(40) 5 or more dark dorsal stripes, at least on nape or anterior body 42
No more than 3 dark dorsal stripes 43

42(41) One row of large prominent white upper lateral spots; all dark dorsal stripes extend length of body ..... ***alacer***
2 or more rows of small white upper lateral spots; most dark dorsal stripes except vertebral present only on nape or fade on forebody ..... ***septenarius***

43(41) Subdigital lamellae narrowly callose to bluntly keeled **(i)** 44
Subdigital lamellae broadly callose to smooth **(c)** 45

44(43) Usually 7 upper labials; nasals usually separated; ne. outcrops and woodlands ..... ***monticola***
Usually 8 upper labials; nasals usually in contact; wstn MGD ..... ***joanae*** (part)

45(43) Upper flanks dark olive with diffuse-edged pale spots 46
Upper flanks black with sharp-edged white spots 47

46(45) Vertebral stripe narrow and often failing to reach hips; upper lateral spots large, squarish and usually confluent with pale midlateral stripe; estn CYP ..... ***nullum***
Vertebral stripe usually broad and extending beyond hips; upper lateral spots small, not contacting midlateral stripe; widespread in Qld ..... ***robustus*** (part)

47(45) Pale paravertebral stripe with black outer edges; white midlateral stripe edged below with a solid black line........ ***eurydice*** (part)
Pale paravertebral stripe, if present, lacking a black outer edge; white midlateral stripe edged below with black and white stippling........ ***arcanus***

48(40) Dark vertebral stripe broad........ ***serotinus***
Dark vertebral stripe narrow........ ***olympicus***

Ctenotus agrestis.
Brendallen Stn, Aramac.

Mitchell Grass Ctenotus

Ctenotus agrestis

SVL 74mm

Stripes and an upper lateral row of pale dashes; broadly-callose subdigital lamellae; inconspicuous rounded ear lobules; single row of supradigital scales with transverse sutures along entire length of 4th toe except base; relatively short tail. Pale greyish brown with indistinctly pale-edged dark vertebral stripe (usually broad and prominent, rarely narrow and obscure) from nape to base of tail; pale dorsolateral stripe from above eye to tail-tip; upper lateral line of pale dashes from eye to hindlimb; pale midlateral stripe from nostril to tail; weak, dark lower lateral and ventrolateral stripes.

Endemic. MGD; deeply cracking clay plains with low scattered acacias from Aramac and Barcaldine, w. to Lochern and Bladensburg NPs.

Lively Ctenotus

Ctenotus alacer

SVL 62mm

Six pale dorsal stripes; black upper flanks enclosing prominent row of white spots; bluntly to narrowly keeled subdigital lamellae; rounded to pointed ear lobules; oblique sutures dividing supradigital scales along basal quarter to half of 4th toe, single scales and transverse sutures along remainder. Black with reddish paravertebral and dorsal stripes, cream dorsolateral stripes, row of large, longitudinally elongate white upper lateral spots, sharp white midlateral stripe on posterior body, often breaking at mid or anterior body.

Rocky areas in far nthn CHC, Cloncurry area in NWH. Also sthn NT, estn WA.

Ctenotus alacer. Alice Springs, NT.

Ctenotus allotropis. Bendidee SF.

Ctenotus arcanus. Mt Mee SF.

Ctenotus arcanus. Pale island colour form. North Stradbroke Is.

Brown-blazed Wedgesnout Ctenotus

Ctenotus allotropis

SVL 55mm

Prominent stripes but no dark vertebral stripe; laterodorsal and upper lateral rows of pale spots; black-keeled subdigital lamellae, blunt at base, finely mucronate towards tip of digit; low rounded ear lobules; nasal scales in short to moderately long contact; single row of supradigital scales with transverse sutures along entire length of 4th toe except base; small size. Brown above, wide black laterodorsal stripe enclosing series of pale spots, narrow white dorsolateral stripe from above eye onto tail, black upper flanks enclosing 2–3 series of pale, vertically aligned spots, prominent white midlateral stripe, dark lower flanks with diffuse white spots.

Dry woodlands, shrublands on reddish sandy soils and loams in BB, ML. Also central NSW.

Arcane Ctenotus

Ctenotus arcanus

SVL 87mm

Variable development of dark vertebral stripe; narrowly callose subdigital lamellae; low rounded to weakly pointed ear lobules; oblique sutures dividing supradigital scales along most of 4th toe from base, single scales and transverse sutures along remainder; large size. On mainland, shades of brown, with prominent pale-edged dark vertebral stripe from nape to base of tail on juv., sometimes contracting to line on nape or disappearing on adults, cream dorsolateral and black laterodorsal stripes, black upper flanks enclosing row of sharp white dots, white midlateral stripe, dappled grey lower flanks, dark-striped limbs. On Stradbroke and Moreton Is., little or no vertebral stripe, smaller, more numerous, pale lateral spots, finely mottled limbs.

Rock outcrops, heaths, woodlands, moist forest edges in SEQ; outlying pop. at Carnarvon NP, BB. Also ne. NSW. Island pops occur in heath on sand.

Ctenotus ariadnae. Nappa Merrie Stn.

Ariadna's Ctenotus
Ctenotus ariadnae
SVL 64mm

Numerous ill-defined stripes, no spots; bluntly keeled to narrowly callose subdigital lamellae; long sharp ear lobules; single row of supradigital scales with transverse sutures along entire length of 4th toe except base. Brown to black with 18–20 very narrow pale stripes, well defined on back, increasingly obscure on flanks. Pale midlateral stripe moderately to well developed, all lateral pattern fragmented in front of forelimb. Sandy spinifex deserts in far wstn CHC. Also nthn SA, central WA.

Ashy Downs Ctenotus
Ctenotus astarte
SVL 82mm

Pattern dominated by pale spots; bluntly keeled to narrowly callose subdigital lamellae; acute to weakly pointed ear lobules; usually single row of supradigital scales with transverse sutures along entire length of 4th toe except base; 37 or fewer

Ctenotus astarte. Morney Stn.

midbody scales; robust build. Greyish brown to yellowish brown, pale spots and flecks often forming irregular, broken paravertebral and dorsolateral lines, small, irregular dark blotches tending to concentrate mainly on mid-dorsal region. Upper flanks have numerous pale dots, tending to align vertically.

Endemic. CHC, n. to Boulia and Diamantina Lakes area, favouring clay and stony soils sparsely vegetated with open shrublands.

Short-clawed Ctenotus
Ctenotus brachyonyx
SVL 83mm

Status uncertain. Falls genetically within an expanded concept of *Ctenotus inornatus*. Reduced stripes, with little or no pale edge to dark vertebral stripe; no spots; broadly callose subdigital lamellae; weakly pointed ear lobules; single row of supradigital scales with transverse sutures along entire length

Ctenotus brachyonyx. Currawinya NP.

of 4th toe except base. Brown to olive with broad black vertebral and laterodorsal stripes ending abruptly at hips or base of tail, pale dorsolateral stripe extending onto tail, sometimes indistinct dark upper lateral dots and pale flecks, obscure pale midlateral stripe.

Dry sandy areas with spinifex, loams supporting chenopod shrublands and mulga woodlands, in ML, far wstn BB, n. to Chesterton Ra. Also NSW, Vic, SA.

Ctenotus brevipes. Undara NP.

Short-footed Ctenotus

Ctenotus brevipes

SVL 60mm

Little or no vertebral stripe; no pale spots; narrowly to broadly callose subdigital lamellae; no ear lobules; single row of supradigital scales with transverse sutures along entire length of 4th toe except base; small size, slender build, short limbs. Brown above with vertebral stripe usually absent (occasionally thin line on nape or foreback), black laterodorsal and white dorsolateral stripes, patternless black upper flanks, prominent white midlateral stripe with dark lower edge, dark lower lateral stripe from ear to behind forelimb.

Endemic. Woodlands of CYP, EIU and GUP. *C. brevipes* exhibits unique characteristics being the only *Ctenotus* lacking ear lobules.

Ctenotus calurus. Craven's Peak.

E. Vanderduys

Blue-tailed Ctenotus

Ctenotus calurus

SVL 49mm

Bold striped pattern and blue tail; finely keeled and mucronate subdigital lamellae; 3–5 ear lobules, 2nd usually largest; large keeled scales on sole opposite 4th toe; single row of supradigital scales with transverse sutures along entire length of 4th toe except base; small size. Black with 8 sharp-edged white stripes; paravertebrals, dorsals, dorsolaterals and midlaterals. Head and foreback flushed with red, and tail with blue.

CHC, in sandy spinifex desert, in vicinity of Ethabuka and Craven's Peak. Also sthn NT and WA.

Capricorn Ctenotus

Ctenotus capricorni

SVL 65mm

Stripes and small, weak, upper lateral spots; moderately

Ctenotus capricorni. Jericho.

broadly callose subdigital lamellae; blunt to weakly pointed ear lobules; single row of supradigital scales with transverse sutures along entire length of 4th toe except base. Olive brown above with pale-edged black vertebral stripe to base of tail, sometimes weak pale brown dorsal stripe to hips, obscure, narrow black laterodorsal stripe to base of tail, pale dorsolateral stripe continuing well onto tail. Upper flanks greyish brown with row of small pale dots or dashes, pale diffuse midlateral stripe.

Endemic. Semi-arid sandy areas with spinifex between about Jericho, DEU and Bladensberg NP, nthn CHC.

D. Knowles

Ctenotus decaneurus. Mueller's Range.

Ten-lined Ctenotus

Ctenotus decaneurus

SVL 50mm

Bold striped pattern, lacking spots; broadly callose subdigital lamellae; small rounded ear lobules; single row of supradigital scales with transverse sutures along entire length of 4th toe except base. Black, usually with 10 sharp-edged pale stripes – paravertebrals, dorsals, dorsolaterals, midlaterals, lower laterals. Limbs marked with bold dark and pale stripes.

Stony hills from NWH to nthn CHC. Also nthn NT and WA.

Ctenotus dux. Ethabuka Stn.

Narrow-lined Ctenotus

Ctenotus dux

SVL 65mm

Numerous stripes, most very narrow but incl. broad, prominent, pale midlateral; no spots; narrowly callose subdigital lamellae; pointed ear lobules; single row of supradigital scales with transverse sutures along entire length of 4th toe except base. Black with 18 or more mostly coppery brown, pale stripes, incl. broad conspicuous white midlateral extending well forward of forelimb. Limbs striped with brown and black.

Desert sand dunes vegetated with spinifex in far wstn CHC. Also deserts of NT, SA, WA.

Brown-backed Yellow-lined Ctenotus

Ctenotus eurydice

SVL 75mm

Prominent pale stripes; usually white upper lateral spots; broadly callose subdigital lamellae; weakly pointed ear lobules; oblique sutures dividing supradigital scales along basal half of 4th toe, with single scales and transverse sutures along remainder. Olive brown above with broad, pale-edged

E. Vanderduys

Ctenotus eurydice. Girraween NP.

black vertebral stripe, black laterodorsal stripes, pale dorsolaterals and black upper flanks with variable white spots (widely spaced row along body, few anteriorly, or none). Prominent white midlateral stripe loops over ear opening and extends forward under eye to snout. Status uncertain, possibly representing a variant of *C. taeniolatus* exhibiting spotted flanks.

Rocky areas, mainly in cool uplands, in NET, SEQ, far estn BB. Also ne. NSW.

Ctenotus eutaenius. Great Basalt Wall.

Black-backed Yellow-lined Ctenotus

Ctenotus eutaenius

SVL 90mm

Prominent stripes; upper lateral stripe often broken anteriorly to form pale dashes; broadly callose subdigital lamellae; weakly pointed ear lobules; oblique sutures dividing supradigital scales along at least basal third of 4th toe, single scales and transverse sutures along remainder. Blackish brown with broad, pale-edged black vertebral stripe, sometimes narrow brown dorsal stripe, black laterodorsal and white dorsolateral stripes, pale upper lateral stripe normally broken into dashes from behind eye for varying distance to about midbody, white midlateral stripe.

Endemic. Tropical woodlands, often assoc. with rock outcrops, in nthn BB, EIU, WT.

Ctenotus hebetior hebetior. Windorah.

Ctenotus hebetior schuettleri. Mary Kathleen Dam.

Five-lined Ctenotus

Ctenotus hebetior

SVL 60mm

Five dark stripes on back; spotted flanks; pale midlateral stripe broken anteriorly; narrow to broadly callose subdigital lamellae; blunt to weakly pointed ear lobules; single row of supradigital scales with transverse sutures along entire length of 4th toe except base. Brown with dark vertebral, dorsal and laterodorsal stripes, pale dorsolateral stripe, dark reddish brown upper flanks enclosing 2 or more series of pale spots or dashes, short pale midlateral stripe, continuous in front of hindlimb and breaking at or before midbody. Ill-defined ssp. *C. h. schuettleri* said to have sharper pattern.

Endemic. Wide variety of semi-arid to arid habitats, from sandy flats and dune-slopes to heavy loams and rocky soils in CHC, MGD, DEU, EIU, GUP (*C. h. hebetior*) and NWH (*C. h. schuettleri*).

Ctenotus helenae. Hay River, NT.

Dusky Ctenotus

Ctenotus helenae

SVL 95mm

Little or no pattern; broadly-callose subdigital lamellae; sharp ear-lobules; single row of supradigital scales with transverse sutures along entire length of 4th toe except base. Brown to olive with pattern completely absent or very weak and reduced to dark vertebral stripe, normally without a pale edge, faint indications of dark laterodorsal and pale dorsolateral stripes, dark lateral flecks and a pale midlateral stripe. Genetic study links this with *C. inornatus*, *C. saxatilis* and *C. brachyonyx*, and subject to confusion with pops tentatively assigned to *C. saxatilis* in sw. Qld. Weakly patterned individuals generally identified as *C. helenae* grading evenly to more prominently striped and spotted specimens treated as *C. saxatilis*.

Spinifex deserts of western CHC, and a doubtful northern record from Mary Kathleen area in NWH. Also NT and WA.

Ctenotus ingrami. Bendidee SF.

Ingram's Ctenotus

Ctenotus ingrami

SVL 84mm

Simple stripes, with white midlateral most prominent; broadly callose subdigital lamellae; weakly pointed ear lobules; oblique sutures dividing supradigital scales along at least basal third of 4th toe, single scales and transverse sutures along remainder. Dark brown above with black vertebral stripe (with or without narrow pale edge) ending on base of tail, black laterodorsal and sharp white dorsolateral stripes from above eye to hips, patternless black upper flanks, very conspicuous broad white midlateral stripe from snout through ear onto tail, white lower lateral stripe.

Dry woodlands on heavy, clay to stony soils in BB, ML, DEU. Also nthn interior of NSW.

R. Valentic

Ctenotus inornatus. Lawn Hill NP.

C. Dollery, QPWS

Ctenotus joanae. Astrebla NP.

Plain Ctenotus

Ctenotus inornatus

SVL 95mm

Extremely variable. Weak pattern but conspicuous pale dorsolateral stripes; broadly callose subdigital lamellae; low to weakly-pointed ear-lobules; single row of supradigital scales with transverse sutures along entire length of 4th toe except base. Pattern strong on juv and weak on adults. Brown with narrowly pale-edged black vertebral stripe reaching base of tail, reduced to a line on nape or absent, and prominent narrow pale dorsolateral stripe with narrow black upper edge. Upper flanks with pale spots on juv, breaking to flecks and mottling on adults. Pale midlateral stripe rarely reaches forward beyond forelimb but sometimes extends to face on juv. Genetic study links this with *C. helenae*, *C. saxatilis* and *C. brachyonyx* and more work needed to determine status of these, but treated here separate spp.

Rock outcrops, river margins and grassy woodlands in NWH and wstn GUP, widely separated from 2 old specimens from 'Near Lockhart River', estn CYP. Also nthn NT to nthn WA.

Blacksoil Ctenotus

Ctenotus joanae

SVL 80mm

Very broad black vertebral stripe; prominent pale dorsolateral and midlateral stripes; broad areas of ground colour; bluntly keeled subdigital lamellae; blunt to weakly pointed ear lobules; oblique sutures normally dividing supradigital scales along at least basal third of 4th toe, with single scales and transverse sutures along remainder. Olive brown with pale-edged black vertebral stripe from nape to base of tail, narrow pale dorsolateral and broad black laterodorsal stripes from behind eye to tail, upper flanks sometimes enclosing a few anterior pale spots, prominent pale midlateral stripe extending forward under eye to snout.

Open grasslands on cracking clay soils in CHC, MGD, wstn GUP. Also estn NT, ne. SA. One of suite of 'black soil' endemics restricted to this featureless habitat, where only available cover is deep, labyrinthine soil cracks.

Ctenotus lateralis. Adels Grove.

Ctenotus leae. Ethabuka Stn.

Gravelly-soil Ctenotus

Ctenotus lateralis

SVL 85mm

Simple striped pattern, lacking spots; broadly callose subdigital lamellae; blunt to weakly pointed ear lobules; single row of supradigital scales with transverse sutures along entire length of 4th toe except base. Brown with prominent stripes – pale-edged black vertebral, black laterodorsal and cream dorsolateral stripes, pale upper lateral stripe, prominent white midlateral stripe.

Heavy, usually stony soils supporting spinifex and woodlands in NWH, GUP, MGD, CHC, EIU.

Centralian Coppertail

Ctenotus leae

SVL 60mm

Simple striped pattern, flushed with red from hips onto tail; finely keeled and mucronate subdigital lamellae; blunt to weakly pointed ear lobules; single row of supradigital scales with transverse sutures along entire length of 4th toe except base. Blackish brown with narrow, sharp-edged pale paravertebral, dorsal and dorsolateral stripes, broad white midlateral stripes.

Crests and slopes of dunes in spinifex deserts in CHC. Also NT, SA, WA.

Ctenotus leonhardii. Quilpie district.

Common Desert Ctenotus

Ctenotus leonhardii

SVL 78mm

Three dark stripes on back; spotted upper flanks; pale midlateral stripe broken anteriorly; narrowly callose subdigital lamellae; blunt to weakly pointed ear lobules; single row of supradigital scales with transverse sutures along entire length of 4th toe except base. Brown above with pale-edged black vertebral stripe, black laterodorsal and pale dorsolateral stripes. Upper flanks dark brown with several series of small pale spots. Pale midlateral stripe present in front of hindlimb, breaking into dashes at midbody. Lower flanks brown with obscure pale spots.

Mainly heavy loams and stony soils in ML, CHC, MGD, NWH, DEU, wstn BB. Also arid zones of all mainland States except Vic. Long-limbed, extremely swift, often venturing into exposed areas.

Ctenotus monticola. Granite Gorge, Mareeba district.

Atherton Ctenotus

Ctenotus monticola

SVL 61mm

Stripes and vertically aligned upper lateral spots; broad area of dorsal ground colour; narrow to moderately broadly callose subdigital lamellae; blunt to weakly pointed ear lobules; oblique sutures dividing supradigital scales along at least basal third of 4th toe, single scales and transverse sutures along remainder. Brown to reddish brown above with narrow black vertebral stripe (with very weak, narrow pale edge), occasionally some indication of narrow dark dorsal stripe, black laterodorsal and narrow white dorsolateral stripes, blackish brown upper flanks enclosing vertically aligned pale spots, narrow white midlateral stripe from hindlimb forward to ear or below eye, dark brown lower flanks with irregular pale spots.

Endemic. EIU, WT, between about Mareeba and Tully. Eucalypt woodlands associated with granite outcrops in w.; presumably exposed rocky sites in e.

Ctenotus nullum. Laura district.

Cooktown Ctenotus

Ctenotus nullum

SVL 79mm

Stripes and squarish pale upper lateral blotches; broadly callose subdigital lamellae; pointed ear lobules; oblique sutures dividing supradigital scales along at least basal third of 4th toe, single scales and transverse sutures along remainder. Olive to reddish brown above with very narrow black vertebral stripe (with weak, narrow pale edge) reduced to line on nape or extending onto back or tail, black laterodorsal and pale dorsolateral stripes, black upper flanks enclosing row of large, squarish pale blotches contacting pale midlateral stripe, grey lower flanks blotched with white.

Endemic. Laura and Cooktown areas, CYP, far nthn EIU. Largely associated with sandstone habitats; also occurs on granite at Black Mtn.

Ctenotus olympicus. Betoota.

Saltbush Ctenotus
Ctenotus olympicus
SVL 69mm

Ragged pattern; subdigital lamellae each with narrow dark brown callus; bluntly pointed ear lobules; oblique sutures dividing supradigital scales along at least basal third of 4th toe, single scales and transverse sutures along remainder. Pale to dark brown above with narrow, dark vertebral stripe (with or without weak, narrow, pale edge) between nape and hips, usually black laterodorsal stripe with inner edges irregular and broken by pale flecks, sometimes cream dorsolateral stripe breaking on body or occasionally reaching hips, black upper flanks enclosing 3–4 rows of small irregular pale spots, midlateral row of pale dashes sometimes forming an irregular posterior stripe.

Shrublands on heavy, often stony soils in CHC. Also arid zones of NT, NSW, SA.

Identification as *C. olympicus* of specimen pictured from Betoota, with a clear pattern incl. sharp row of pale laterodorsal spots, is tentative.

Leopard Ctenotus
Ctenotus pantherinus
SVL 94mm

No stripes, but rows of dark-edged pale spots; finely keeled to mucronate subdigital lamellae; blunt to weakly pointed ear lobules; single row of supradigital scales with transverse sutures along entire length of 4th toe except base; large size, robust build. Coppery brown, boldly marked with white longitudinal dashes, each highlighted by black edge.

Arid areas with spinifex e. to Alton NP. Qld pop. referred to as *C. p. acripes*, occurs through NT to nw. coast of WA. Other ssp. in arid parts of all mainland States except Vic.

Pianka's Ctenotus
Ctenotus piankai
SVL 60mm

Simple striped pattern; bluntly keeled to narrowly callose subdigital lamellae; 1–5 acute ear lobules; single row of supradigital scales with transverse sutures along entire length of 4th toe except base. Brown to reddish brown with 6 (occasionally 8) pale stripes; paravertebrals, dorsolaterals, midlateral and occasionally ventrolaterals.

Sand dunes with spinifex at Craven's Peak, CHC. Also sandy areas with spinifex in NT and WA.

E. Vanderduys

Ctenotus piankai. Craven's Peak.

Ctenotus pantherinus. Ethabuka Stn.

R. Browne-Cooper

Ctenotus pulchellus. Mt Isa.

Pretty Ctenotus

Ctenotus pulchellus

SVL 85mm

Five dark dorsal stripes; little or no pale midlateral stripe but sharply contrasting upper and lower flanks; narrowly to broadly callose subdigital lamellae; blunt to pointed ear lobules; single row of supradigital scales with transverse sutures along entire length of 4th toe except base. Reddish brown above with black vertebral, dorsal and laterodorsal stripes, white dorsolateral stripe, blackish brown upper flanks, burnt orange mid to lower flanks, dense white spotting over all lateral surfaces.

Heavy red soils, stony hills with spinifex in NWH, MGD. Also ne. NT.

E. Vanderduys

Ctenotus quinkan. Jowalbinna Stn.

Quinkan Ctenotus

Ctenotus quinkan

SVL 80mm

No dorsal pattern; moderately callose subdigital lamellae; sharp ear lobules; oblique sutures dividing supradigital scales along almost entire length of 4th toe. Coppery brown above, cream dorsolateral stripes, patternless black flanks and white to pale pink midlateral stripe with black lower edge extending forward to ear.

Endemic. Restricted to sandstone habitats of estn CYP, from n. of Cooktown to Laura area.

S. Donnellan

Ctenotus rawlinsoni. Cape Flattery.

Cape Heath Ctenotus

Ctenotus rawlinsoni

SVL 80mm

Prominent stripes and rarely spots; moderately callose subdigital lamellae; short, pointed ear lobules; oblique sutures dividing supradigital scales along almost entire length of 4th toe. Black with conspicuous pale paravertebral and dorsolateral stripes extending well onto tail, patternless black upper flanks (rarely a few anterior pale spots), and broad white midlateral stripe extending forward to ear, edged below with dark stripe and breaking into blotches and bars between forelimb and ear.

Endemic. Confined to heaths on white coastal sands in vicinity of Cape Bedford and Cape Flattery, CYP.

Royal Ctenotus
Ctenotus regius
SVL 73mm

Three dark stripes on back; spotted upper flanks; pale midlateral stripe extends well forward of forelimb; broadly keeled to narrowly callose subdigital lamellae; blunt (juv.) to pointed (adult) ear lobules; normally single row of supradigital scales with transverse sutures along entire length of 4th toe except base, rarely oblique sutures on basal quarter to third. Brown above, with sharply pale-edged, narrow black vertebral stripe, black laterodorsal and pale dorsolateral stripes, dark reddish brown to black upper flanks enclosing scattered pale spots or dashes, unbroken white midlateral stripe reaching forward to ear or upper lip, brown lower flanks with scattered pale spots.

Red soils, often in association with spinifex, in semi-arid to arid CHC, ML, wstn BB. Also arid areas of all mainland States.

Ctenotus regius. St George.

Eastern Striped Skink
Ctenotus robustus
SVL 123mm

Status and available name uncertain. Genetic study links this with *C. spaldingi* and restricts true *C. robustus* to nthn NT and WA. Normally three dark stripes on back; pale upper lateral blotches; 4 supraocular scales; smooth to broadly-callose subdigital lamellae; weakly pointed ear lobules; oblique sutures dividing supradigital scales along at least basal third to half of 4th toe, single scales and transverse sutures along remainder. Brown to olive with broad, prominent, pale-edged black vertebral stripe, black laterodorsal and prominent pale dorsolateral stripes, dark olive brown to dark brown upper flanks with diffuse, pale, upper lateral blotches (sharp-edged spots on juv.), prominent pale midlateral stripe reaching forward at least to ear and represented on upper lip as pale line under eye. Pops from sandy south coastal areas, incl. is. of Moreton Bay, range from boldly marked to fawn and completely patternless.

Ctenotus robustus. Wynnum.

Ctenotus robustus. Pale island form. North Stradbroke Is.

Very widespread, possibly excepting GUP, NWH and CHC, in habitats ranging from woodlands, heaths and rock outcrops to weedy suburban sites. Also NSW, Vic and SA.

A. Kutt

Ctenotus rosarium. Fortuna Stn.

Beaded Ctenotus

Ctenotus rosarium

SVL 43mm

Weak dorsal and strong upper lateral pattern; narrowly callose subdigital lamellae; weakly pointed ear lobules; single row of supradigital scales with transverse sutures along entire length of 4th toe except base. Copper brown above, very narrow, finely pale-edged black vertebral stripe extending for varying distance to base of tail, very narrow dark laterodorsal and pale dorsolateral stripes, blackish brown upper flanks with row of prominent, round whitish spots, prominent white midlateral stripe extending forward to between eye and snout, dark lower lateral stripe enclosing diffuse anterior pale blotches.

Endemic. Restricted to narrow band of red sandy soils with spinifex-dominated open woodlands from just n. of Aramac to White Mtns NP in DEU.

Ctenotus saxatilis. Weakly patterned. Nappa Merrie Stn.

Ctenotus saxatilis. Mt Isa.

Ctenotus saxatilis. Nappa Merrie Stn.

Rock Ctenotus

Ctenotus saxatilis

SVL 100mm

Stripes and upper lateral spots; no pale stripe under eye; broadly callose subdigital lamellae; 8 upper labial scales; weakly pointed ear-lobules; single row of supradigital scales with transverse sutures along entire length of 4th toe except base. Olive brown to reddish brown above with strong to weak pattern; pale-edged dark vertebral stripe, dark laterodorsal and white dorsolateral stripes, brown to blackish brown upper flanks with pale blotches or flecks, and pale midlateral stripe extending forward to about forelimb. Genetic study links this with *C. helenae*, *C. inornatus* and *C. brachyonyx* and more work needed to determine status of these. Treated here separate spp. and identified as saxatilis with a degree of uncertainty; the strongly marked specimens are treated here as *C. saxatilis*, but there is a gradation in sw. deserts to sympatric weakly patterned individuals often identified as *C. helenae*.

Variety of habitats from stony slopes and sandy flats with spinifex in CHC to pandanus-lined creeks in NWH and GUP. Also NT, SA and WA.

H. Cogger

Ctenotus schevilli. Muttaburra area.

Spotted Black-soil Ctenotus

Ctenotus schevilli

SVL 85mm

Pattern dominated by pale spots; bluntly keeled to narrowly callose subdigital lamellae; weakly pointed ear lobules; single row of supradigital scales with transverse sutures along entire length of 4th toe except base; 40 or more midbody scale rows; robust build. Olive brown above with pale reddish brown lateral flush, dark dorsal blotches sometimes coalescing to form irregular vertebral stripe, numerous small pale spots scattered over back and arranged in roughly vertical series on flanks.

Endemic. MGD. Deeply cracking grassy clay plains in Richmond, Muttaburra, Aramac districts.

Ctenotus schomburgkii. Windorah.

Barred Wedge-snouted Ctenotus

Ctenotus schomburgkii

SVL 52mm

Five dark dorsal stripes; large reddish brown upper lateral blotches; finely keeled and mucronate subdigital lamellae; nasal scales usually separated; low rounded ear lobules; single row of supradigital scales with transverse sutures along entire length of 4th toe except base. Reddish brown to olive with black vertebral, dorsal and laterodorsal stripes, cream dorsolateral stripe, black upper flanks enclosing large, usually squarish, prominent blotches, white midlateral stripe.

Arid sandy areas, incl. shrublands, spinifex grasslands, in ML, CHC. Also arid zones in all mainland States except Vic.

Gibber Ctenotus

Ctenotus septenarius

SVL 72mm

Dorsal pattern of 5–7 dark stripes best developed anteriorly; reddish brown posterior flush; narrowly callose subdigital lamellae; low rounded ear lobules; oblique sutures

Ctenotus septenarius. South Galway Stn.

dividing supradigital scales along at least basal third of 4th toe, single scales and transverse sutures along remainder; very long limbs. Rich reddish brown on mid to posterior back. Nape and foreback dominated by fine, fragmented black and white stripes tending to fade and further fragment posteriorly, with sometimes only the vertebral reaching posterior body or base of tail. Upper flanks black, enclosing prominent white spots. Pale midlateral stripe may extend forward to snout as unbroken stripe or series of spots.

Weathered stony slopes and mesas in CHC. Also NT, SA, WA. Extremely swift, often foraging away from vegtn on open terrain.

Gravel-downs Ctenotus
Ctenotus serotinus
SVL 50mm

Pattern dominated by pale spots and dark vertebral stripe; smooth to bluntly keeled subdigital lamellae; small rounded ear lobules; oblique sutures dividing supradigital scales along at least basal third of 4th toe, single scales and transverse sutures along remainder. Olive brown with diffuse pattern – ragged, weakly pale-edged black vertebral stripe from nape to base of tail, diffuse pale dorsolateral and midlateral stripes, dark olive brown upper flanks. Pale spots sparsely scattered over back, tending to align vertically on flanks.

Endemic. Recorded on dunes, adj. stony soils in Diamantina Lakes area, CHC.

D. Knowles
Ctenotus serotinus. Diamantina Lakes.

Ctenotus spaldingi. Chillagoe.

Spalding's Ctenotus
Ctenotus spaldingi
SVL 100mm

Genetic study links this with sp. tentatively referred to as *C. robustus*. Regarded here as having: only 3 supraocular scales; sharp brow; stripes and pale upper lateral blotches; round to pointed ear lobules; broadly callose subdigital lamellae; oblique sutures dividing supradigital scales along at least basal third to half of 4th toe, single scales and transverse sutures along remainder. Brown to reddish brown above, pale-edged dark vertebral stripe (broad and prominent, reduced to line on nape, or very rarely absent), narrow black laterodorsal and white dorsolateral stripes, dark brown to black upper flanks, upper lateral series of pale squarish blotches, pale midlateral stripe.

Rock outcrops, dry timbered areas, coastal dunes of NWH, GUP, EIU, DEU, WT, CYP, n. through Torres Strait to sthn NG. Also nthn NT.

Ctenotus strauchii strauchii. Belyando.

Ctenotus strauchii varius. Durham Downs Stn.

Eastern Barred Wedge-snouted Ctenotus

Ctenotus strauchii

SVL 55mm

Pale laterodorsal spots; very small, blunt ear lobules; finely keeled and mucronate subdigital lamellae; single row of supradigital scales with transverse sutures along entire length of 4th toe except base. Brown, with or without pale-edged black vertebral stripe from nape to base of tail, black laterodorsal stripe enclosing or partly enclosing pale spots or dashes, white dorsolateral stripe, black upper flanks with vertically aligned pale dots, white midlateral stripe reaching forward to upper lip. Ill-defined ssp. *C. s. varius* said to have more diffuse pattern, paler colour, blunt to weakly pointed ear lobules with upper 1st or 2nd much larger than others.

Woodlands and shrublands, mainly on heavy, often stony soils in BB, ML, MGD, DEU, EIU (*C. s. strauchii*) and CHC (*C. s. varius*). Also adj. NSW, SA, NT.

Ctenotus striaticeps. Mt Isa district.

Stripe-headed Ctenotus

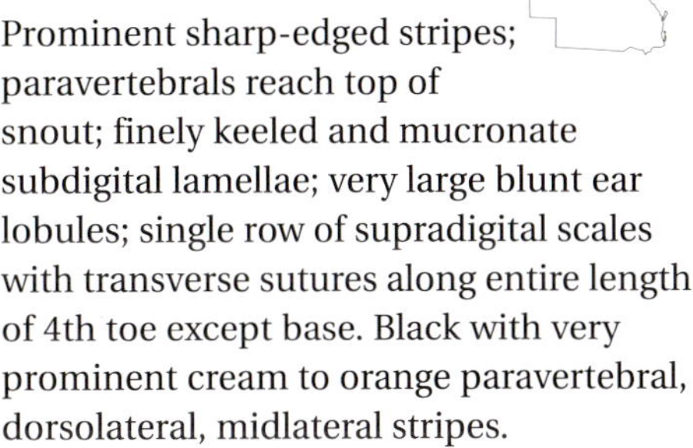

Ctenotus striaticeps

SVL 50mm

Prominent sharp-edged stripes; paravertebrals reach top of snout; finely keeled and mucronate subdigital lamellae; very large blunt ear lobules; single row of supradigital scales with transverse sutures along entire length of 4th toe except base. Black with very prominent cream to orange paravertebral, dorsolateral, midlateral stripes.

Hard stony soils with spinifex in NWH, GUP. Also adj. NT.

Eyrean Ctenotus

Ctenotus taeniatus

SVL 56mm

Obscure, fragmented pattern; finely keeled and mucronate subdigital lamellae; blunt to pointed ear lobules; large keeled scales on sole opposite 4th toe; single row of supradigital scales with transverse sutures along entire length of 4th toe except base; small size. Reddish brown to fawn with very narrow, often broken, dark

Ctenotus taeniatus. Durrie Stn.

vertebral stripe, broken wavy dorsal lines composed of small, dark, longitudinally aligned spots, upper lateral series of dark flecks tending to cluster into vertical bars, sometimes obscure indication of pale midlateral stripe.

Crests and slopes of dunes in CHC. Also adj. NT, SA, NSW and nw. Vic.

Ctenotus taeniolatus. Coominya.

Copper-tailed Skink

Ctenotus taeniolatus

SVL 80mm

Prominent sharp-edged stripes; usually pale streak or 1–2 white spots on temple; small rounded ear lobules; narrowly to bluntly keeled subdigital lamellae; oblique sutures dividing supradigital scales along at least basal third to half of 4th toe, single scales and transverse sutures along remainder. Prominent, pale-edged black vertebral stripe, brown to reddish brown dorsal colour (a broad expanse on back or reduced to narrow stripe), black laterodorsal and pale dorsolateral stripes, black upper flanks, white midlateral stripe extending forward over ear to upper lip, white lower lateral stripe with narrow black lower edge.

Various habitats, often featuring heath understorey on sandy to stony soils, in NET, SEQ, BB, CQC, EIU, WT. Also estn NSW, estn Vic.

Ctenotus terrareginae. Hinchinbrook Is.

Hinchinbrook Ctenotus

Ctenotus terrareginae

SVL 90mm

Stripes but little or no vertebral stripe; broadly callose subdigital lamellae; blunt ear lobules; oblique sutures dividing supradigital scales along at least basal third to two-thirds of 4th toe, single scales and transverse sutures along remainder; very long tail. Dark brown above with vertebral stripe reduced to narrow line on nape and foreback, narrow black laterodorsal and white dorsolateral stripes, patternless black upper flanks, white midlateral stripe extending forward through ear to snout, lower flanks marbled anteriorly with black and white, coalescing to form dark lower lateral stripe on posterior body.

Endemic. Rocky areas of WT, between Paluma area and Hinchinbrook Is.

Little Zebra Ctenotus
Ctenotus zebrilla
SVL 40mm

Prominent stripes and few or no spots; finely dark-keeled and mucronate subdigital lamellae; very small, round to weakly pointed ear lobules; single row of supradigital scales with transverse sutures along entire length of 4th toe except base. Black to blackish brown with narrow pale brown paravertebral and dorsal stripes, broader white dorsolateral stripe, occasionally an upper lateral row of widely spaced small white spots, broad white midlateral stripe extending forward onto upper lip.

Ctenotus zebrilla. Porcupine Gorge.

Endemic. Rocky hills with dry woodlands, low grasses in GUP, EIU, between about Croydon, Lappa, Porcupine Gorge.

Slender Blue-tongues and Pink-tongue Genus *CYCLODOMORPHUS*

9 Aust. sp.; 3 occur in Qld. Long bodies and relatively short, widely spaced limbs each with 5 digits; 3rd and 4th toes roughly equal in length; movable lower eyelid with no transparent disc; ear opening usually with small anterior lobules; widely separated parietal scales; divided frontoparietals; no row of subocular scales between eye and upper labials.

One large, semi-arboreal sp., the Pink-tongue (*C. gerrardii*), is associated with the moist e. while 2 much smaller, wholly terrestrial skinks, the Slender Blue-tongues, occupy arid wstn regions. These relatives of the Blue-tongues (*Tiliqua spp.*) are omnivorous livebearers. They flicker their thick fleshy tongues in apparent mimicry of snakes when alarmed.

KEY TO *CYCLODOMORPHUS*

1	Head broad, distinct from narrow neck; tongue of adult pink	***gerrardii***
	Head narrow, not distinct from neck; tongue of adult blue	2
2	Dark bars on neck	***venustus***
	No dark bars on neck	***melanops***

Pink-tongued Skink
Cyclodomorphus gerrardii
SVL 200mm

Broad head distinct from neck; very long prehensile tail; relatively long, slender digits; forelimbs and hindlimbs nearly meet at midbody; tongue pink on most adults but blue on juv.; large size. Bands present or absent on adults, always present and prominent on juv. Banded forms have prominent dark bands between nape and tail-tip. Unbanded forms plain tan with dark-tipped snouts.

Humid areas in SEQ, BB, CQC, WT, possibly estn EIU, n. to Cairns area. Also estn NSW. Mainly associated with moist timbered habitats; also thrives in well-watered gardens, even in central Brisbane. Diurnal and nocturnal. Feeds mainly on slugs and snails, employing large, rounded rear teeth to crush shells. Skilled climber, recorded foraging in low trees, in ceilings and on curtain rails.

Cyclodomorphus gerrardii. Plain form. Mt Glorious.

Cyclodomorphus gerrardii. Banded form. Mt Glorious.

Spinifex Slender Blue-tongue
Cyclodomorphus melanops
SVL 132mm

Head not distinct from neck; moderately short, thick tail with 58–77 subcaudal scales; very short legs, failing to meet by wide margin; no dark neck blotches; small size. Brown to olive with prominent pale spots on juveniles, pattern on adults reduced to obscure dark edges to scales. Isolated pop from near Amby has prominent scattered dark flecks.

Arid spinifex dominated areas in MGD, NWH and CCH, and open grassland in sthn BB. Represented in Qld by the ssp. *C. m. elongatus*. This and other ssp. also occur in arid parts of all mainland states except Vic.

Cyclodomorphus melanops. Amby area.

Cyclodomorphus melanops. Lark Quarry district.

Saltbush Slender Blue-tongue

Cyclodomorphus venustus

SVL 101mm

Head not distinct from neck; short, thick tail with 44–54 subcaudal scales; very short legs, failing to meet by wide margin; prominent dark neck blotches; small size. Brown to olive with 3–4 vertical black bars on side of neck.

CHC, ML. Recorded from chenopod shrubs on soft sand at Kahmoo Stn near Cunnamulla and stony plain at Betoota. Also nw. NSW, wstn SA.

Cyclodomorphus venustus. Kahmoo Stn.

Genus *EGERNIA*

18 Aust. spp.; 6 occur in Qld. Widely separated parietal scales; well-developed limbs with 5 digits including a 4th toe much longer than 3rd; movable lower eyelid with no transparent disc, and no distinctive pale margins; ear-opening with obvious anterior lobules; medium to extremely large size; robust build; carinated to extremely spiny body scales.

Throughout Qld. Included are some of the world's largest skinks. Some occupy burrows under rocks or logs, others live in narrow rock or wood crevices, while those with spines, wedge themselves so securely that little can budge them. They are omnivorous, and large species eat very significant amounts of vegetation. Livebearers. The young often live with adults in small colonies. Many share communal defecation sites. Diurnal.

KEY TO QUEENSLAND *EGERNIA*

1 Prominent spines on tail 2
No spines on tail 4

2(1) Tail very short and dorsally depressed .... ***stokesii***
Tail moderate to long, and cylindrical 3

a b

3(2) Dorsal scales each with one large spine .... ***cunninghami***
Dorsal scales each with more than one (normally 3–4) spines .... ***hosmeri***

4(1) Scales on parietal region asymmetrical and fragmented (**a**) .... ***rugosa***
Scales on parietal region symmetrical and unfragmented (**b**) .. 5

5(4) Dorsal colour mainly brown; ventral colour mainly bright orange to orange-yellow; uplands of Main Range and Granite Belt .... ***mcpheei***
Dorsal colour mainly grey; ventral colour dull orange-yellow to yellow; dry woodlands of estn interior and a coastal population in Brisbane .... ***striolata***

Egernia cunninghami. Timber-inhabiting. Goomburra, Main Range NP.

Egernia cunninghami. Timber- and rock-inhabiting. Sundown NP.

Egernia cunninghami. Rock-inhabiting Girraween NP.

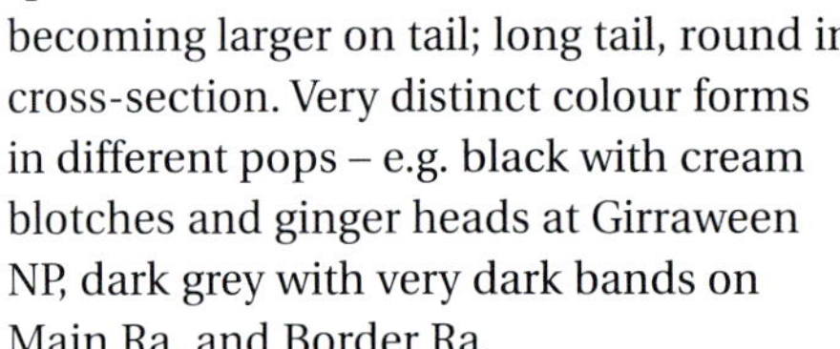

Cunningham's Skink

Egernia cunninghami

SVL 200mm

Very spiny, incl. single large spine on each dorsal scale, becoming larger on tail; long tail, round in cross-section. Very distinct colour forms in different pops – e.g. black with cream blotches and ginger heads at Girraween NP, dark grey with very dark bands on Main Ra. and Border Ra.

NET, SEQ, BB, incl. isolated nthn outlier in Mt Moffatt sector of Carnarvon NP. Also estn NSW, Vic, se. SA. Some distinctive pops mainly rock-inhabiting; others dwell largely in fallen hollow timber.

Egernia hosmeri. Dajarra area.

Hosmer's Skink

Egernia hosmeri

SVL 180mm

Very spiny, incl. 3–4 sharp keels on each dorsal scale, reducing to one long spine on each scale along relatively short, cylindrical tail. Brown with large, dark-edged pale blotches on head, neck and forebody, becoming weaker posteriorly. Pop. from Basalt Wall n. of Charters Towers very dark and virtually patternless.

Rock outcrops in NWH, EIU; possibly nthn DEU. Also ne. NT. Colonies inhabit rock crevices, occasionally hollow timber.

Egernia mcpheei. Girraween NP.

Eastern Crevice-skink

Egernia mcpheei

SVL 143mm

Relatively flat head and body; robust build; blunt keels on dorsal scales of adults; series of enlarged scales along top of tail; sombre colouration. Brown above with narrow, broken dark lines, darker on upper flanks with conspicuous pale marks on upper lips, bright orange to orange-yellow below.

Rainforests, wet and dry sclerophyll forests, rock outcrops in upland areas of NET, sthn BB, SEQ, n. to Main Ra. Also ne. NSW. Shelters in narrow rock and timber crevices. Commonly inhabits exfoliating granite in Girraween NP and rocky fissures beside tracks at Cunningham's Gap.

Egernia rugosa. Alton NP area.

Yakka Skink

Egernia rugosa
SVL 200mm

Fragmented parietal scales; several blunt keels on each dorsal scale; very large, flat ear lobules; very large size; robust build and thick tail. Brown with conspicuously darker brown to black mid-dorsal zone and pale brown dorsolateral stripes.

Endemic. Dry open forests, woodlands, rocky areas of BB, nthn SEQ, EIU, sthn CYP. Lives in communal burrow complexes, often under heaped dead timber, and in deep rock crevices. Often utilises abandoned rabbit warrens. Also recorded under shearing sheds and other rural buildings. Seldom ventures far from cover, often betraying its presence by communal defecation site.

Egernia stokesii. Aramac area.

Stokes' Skink; Gidgee Skink

Egernia stokesii
SVL 180mm

Very spiny, incl. 2 sharp spines on each dorsal scale; short flat tail with one long spine on each scale. Brown with dark and pale scales forming small blotches, arranged in irregular transverse rows.

Dry, timbered and rocky areas of CHC, MGD, DEU. Also arid zones of all mainland States except Vic. Like other spiny *Egernia*, uses its spines to wedge itself firmly and inextricably into wood hollows and rock crevices. Represented in Qld, NSW, SA by ssp. *E. s. zellingi.* Other ssp. occur in WA.

Egernia striolata. Lonesome NP.

Tree Skink

Egernia striolata
SVL 119mm

Relatively flat head and body; robust build; blunt keels on dorsal scales of adults; series of very broad scales along top of tail; sombre colouration. Dark olive to grey with rectangular black blotches concentrated on nape and back, excluding broad pale flush from each side of neck along shoulders to midbody or base of tail, black anterior flanks, conspicuously white to pale grey upper lips and pale orange to dull yellow ventral surfaces, barred with dark pigment on chin and throat.

Dry sclerophyll forests, woodlands in SEQ, BB, NET, ML, MGD, DEU, EIU, mainly w. of coastal ra. but reaching coast near Brisbane with records incl. Boondall, Kippa Ring and Whiteside. Also NSW, Vic, SA. Shelters in hollow timber, beneath loose bark, in narrow rock crevices.

Genus *EMOIA*

About 100+ spp. from estn Pacific to South-East Asia, Japan and Taiwan (China); 2 Aust. spp., both occur in Qld. Well-developed limbs, each with 5 digits; long tails; long snouts; movable lower eyelid enclosing transparent disc; frontoparietal scales fused to form one shield; pair of supranasal scales present; parietal scales in contact; smooth body scales.

Restricted in Qld to nthn CYP and Torres Strait islands, essentially the sthn edge of a broad tropical distribution encompassing many islands, atolls and, to a lesser extent, continents. *Emoia* are efficient rafters, having dispersed on floating debris to colonise remote oceanic landmasses. Diurnal. Egglayers (clutches of 2 eggs).

Mangrove Skink
Emoia atrocostata
SVL 85mm

Sharp to moderate contrast between upper and lateral surfaces; 36 or fewer subdigital lamellae under 4th toe; 61 or more paravertebral scales between nape and rear edge of hindlimb. Brown to greyish brown above with dark and pale spots or flecks. Flanks black with white spots.

Emoia atrocostata. Boigu Is.

Intertidal areas on tip of CYP, and Torres Strait is. Also is. of wstn Pacific, through NG to Japan, Taiwan (China). Coastal rocks and mangroves, foraging within splash zone, where they are displaced by rising tides. Marine invertebrates such as polychaete worms included in diet. Strong swimmers, readily taking to water if pursued.

Long-tailed Skink
Emoia longicauda
SVL 98mm

Very long snout; little or no contrast between upper and lateral surfaces; 63 or more subdigital lamellae under 4th toe; 58 or fewer paravertebral scales between nape and rear edge of hindlimb. Brown, either uniform or with few scattered pale flecks on back and upper flanks, grading to bright lemon yellow or yellowish green on lips, lower flanks and ventral surfaces.

Nthn CYP, s. to McIlwraith Ra. and n. through is. of Torres Strait to NG and Solomon Is. Monsoon forests, clearings and edges, incl. gardens, plantations in some nthn areas. Arboreal, basking on branches and vertical trunks.

Emoia longicauda. Moa Is.

Sand-swimmers
Genus *EREMIASCINCUS*

11 Aust spp.; 4 others in estn Indonesia and Timor; 6 in Qld. Well-developed limbs, each with 5 digits; movable lower eyelid lacking a transparent disc; ear opening present and conspicuous without lobules; parietal scales in contact; frontoparietals divided; glossy body scales; pattern comprises dark flecks tending to concentrate on upper lateral zone, or simple dark bands.

Sandy deserts to nthn river margins and monsoon forests. Also tropical to arid zones in NT, WA and SA. Crepuscular to nocturnal, sheltering by day in burrows (including those of rabbits), soil cracks, rocks, logs and under thick leaf-litter. Egglayers.

KEY TO *EREMIASCINCUS*

1 Pattern predominantly banded, at least on tail; dorsal scales on posterior body, hips and/or tail with low ridges 2
Pattern dominated by flecks, often most concentrated on upper lateral zone; scales on posterior body and hips flat with no ridges 5

a

2(1) Last upper labial scale divided horizontally; a row of scales (subinfralabials) between lower labials and chin shields (**a**) .... ***fasciolatus***
Last upper labial scale entire; no row of scales between lower labials and chin shields 3

3(2) Upper labial scales 7 .... ***intermedius***
Upper labial scales 8 4

4(3) Bands prominent and broad, covering 2 or more scale rows on body .... ***richardsonii***
Bands weak, often absent on body, covering less than 2 scale rows .... ***phantasmus***

5(1) Paravertebral scales significantly broader than adjacent scales; adpressed limbs widely separated .... ***pardalis***
Paravertebral scales roughly equal to adjacent scales; adpressed limbs overlap or nearly overlap ... ***isolepis***

Eremiascincus fasciolatus. Glenmorgan.

R. Valentic

Eremiascincus intermedius. Mt Isa area.

Eastern Narrow-banded Skink

Eremiascincus fasciolatus

SVL 123mm

Banded; one lower labial scale contacts postmental scale; subinfralabial scales present; usually 8 upper labial scales; last upper labial scale divided horizontally; scales along top of 4th toe have oblique sutures along almost entire digit; dorsal ridges present on posterior body and tail; large size and robust build. Pale yellowish brown to reddish brown with 10–16 narrow bands, no more than one scale wide on body, and 25–35 very narrow bands on original tail.

Endemic. SEQ, BB, CQC and DEU, in open subtropical forests, semiarid woodlands and heathlands inland to Roma and Moonie areas, s. to Purga and Wivenhoe Dam, reaching coast at Rockhampton, Port Curtis and Curtis Is., and n. to Mt Cooper and Blackbraes NP.

Northern Narrow-banded Skink

Eremiascincus intermedius

SVL 88mm

Banded; one lower labial scale broadly contacts postmental scale; subinfralabial scales absent; 7 upper labial scales; last upper labial scale entire; scales along top of 4th toe have transverse sutures along at least distal third of digit; dorsal ridges present on posterior body and tail. Brown to reddish brown with 6–16 transverse narrow bands, less that 2 scales wide, on body and up to 42 on original tail.

Spinifex grasslands in NWH. Also arid and to lesser extent tropical NT.

Eremiascincus isolepis. Undara NP.

Northern Bar-lipped Skink

Eremiascincus isolepis

SVL 72mm

Pattern dominated by flecks; one lower labial scale contacts postmental scale; subinfralabials absent; scales on posterior body and hips flat with no ridges; limbs separated by about one forelimb length or less when pressed to sides. Brown, lightly to heavily spotted with darker brown to black. Spots are scattered or tend to form longitudinal rows on back, cluster on upper flanks and become sparser on mid to lower flanks.

Often associated with moist pockets within dry woodlands and outcrops, but also frequently found in floodplains and margins of swamps in NWH, GUP and EIU. Also nthn NT to nthn WA.

Lowlands Bar-lipped Skink

Eremiascincus pardalis

SVL 70mm

Pattern dominated by flecks; one lower labial scale contacts postmental scale; subinfralabials absent; scales on posterior body and hips flat with no ridges; limbs separated by about one forelimb length or a little more when pressed to sides. Brown with numerous dark flecks. These may be absent, scattered, or form a pair of paravertebral rows on back, and tend to concentrate to form dense black spots on dorsolateral area and flanks.

Endemic. Woodlands, monsoon forests and vine thickets of WT, northern EIU, CYP and eastern GUP.

E. Vanderduys

Eremiascincus pardalis. Pormpuraaw.

Eremiascincus richardsonii. Currawinya NP.

Eremiascincus phantasmus. Nappa Nerrie Stn.

Ghost Skink

Eremiascincus phantasmus

SVL 92mm

Weakly banded; two lower labial scales contact postmental scale; subinfralabial scales absent; usually 8 upper labial scales; last upper labial scale entire; scales along top of 4th toe have transverse sutures along half to almost entire digit; dorsal ridges present on posterior body and tail. Pale pinkish brown to pale grey with 29–39 very narrow bands on tail. Bands on body often absent; narrow, weak and largely confined to flanks when present.

Sand dunes with spinifex and cane grass in CHC. Also sandy deserts of ne. SA and sthn NT.

Broad-banded Sand-swimmer

Eremiascincus richardsonii

SVL 113mm

Banded; one lower labial scale contacts postmental scale; subinfralabial scales absent; 8 upper labial scales; last upper labial scale entire; scales along top of 4th toe have transverse sutures along about distal third of digit; dorsal ridges present on posterior body and tail. Pink to reddish brown with dark bands relatively few, broad and often oblique; 8–14 between nape and hips and 19–32 on tail.

Arid zones of CHC, ML, MGD, BB, GUP and NWH. Within the desert habitats it probably fares best on heavy and stony soils.

Genus *EROTICOSCINCUS* Monotypic genus.

Elf Skink

Eroticoscincus graciloides

SVL 32mm

Short, well-developed limbs with 4 fingers and 5 toes; large eyes; movable lower eyelid enclosing transparent disc; pointed snout; parietal scales in contact; minute circular ear opening; very small size. Brown (with multi-hued sheen from some angles) with narrow, weak, dark dorsolateral line on body, rufous dorsolateral stripe on tail, obscure dark M-shape on rear of head.

Endemic. Humid areas of SEQ, on ra. and coastal plain between Mt Nebo and Fraser Is. Extremely secretive. Forages in shaded, semi-concealed situations; shelters under logs, leaf litter and other ground debris in vine thickets, wet sclerophyll forests, rainforests and damp depressions in dry sclerophyll forests. Egglayer.

Eroticoscincus graciloides. Tewantin.

Genus *EUGONGYLUS*

3 NG and wstn Pacific spp.; 1 Aust. species occurs in Qld.

Brown Sheen-skink

Eugongylus rufescens

SVL 169mm

2 supranasal scales; movable lower eyelid lacking transparent disc; ear openings with small anterior lobules; parietal scales in contact; frontoparietal scales divided; short, well-developed limbs with 5 digits; long body; large size. Juv. prominently patterned with widely spaced, narrow pale bands. Adults lose all or most pattern and are reddish brown to grey, from some angles reflecting iridescent green/purple sheen.

Monsoon forests, other densely vegetated areas from tip of CYP, through Torres Strait is. to NG. Crepuscular to nocturnal, sheltering by day in and under rotting logs, in tree buttresses and low tree hollows. Lays clutches of up to 4 eggs.

Eugongylus rufescens. Lockerbie Scrub.

Genus *EULAMPRUS*

5 Aust. species; 2 in Qld. Well-developed limbs overlap when pressed to side of body, all with 5 digits; subdigital lamellae longitudinally grooved; 2–3 lower labial scales in contact with postmental scale; scales along top of 4th toe with transverse sutures along length of digit; movable lower eyelid lacking a transparent disc; no lobules on ear opening; parietal scales in contact; frontoparietals divided; smooth, shiny scales.

Moist estn areas. Also estn NSW, Vic, se. SA. One sp. largely associated with waterside environments incl. suburban creeks. The other restricted to high altitudes. Diurnal livebearers. Diets ranges from arthropods to smaller skinks.

Alpine Water Skink

Eulamprus kosciuskoi

SVL 85mm

Coppery brown with weak dark vertebral and laterodorsal stripes, pale dorsolateral stripes, black flanks enclosing several rows of white to yellow spots, pale yellow to grey lower flanks and black spots across belly.

Records from Stanthorpe area, NET. Also alpine areas of NSW and Vic. Woodlands, heaths and tussock grasslands, incl. alpine bogs.

Eulamprus kosciuskoi. New England NP, NSW.

Eastern Water Skink

Eulamprus quoyii

SVL 115mm

Gold dorsolateral stripe; long limbs and tail; large size. Brown to golden brown with sparse black dorsal flecks or blotches, prominent pale dorsolateral stripes from above eye to midbody or hips, black upper flanks enclosing prominent pale spots.

Mainly associated with margins of waterways in NET, SEQ, BB, CQC, WT, n. to Cairns district. Also NSW, se. SA, nw. Vic. Basks conspicuously along water's edge and patrols logs, rocks and other exposed sites. Common in thickly vegetated garden edges in suburban areas with year-round watering, away from permanent water bodies. Formidable predator of smaller skinks.

Eulamprus quoyii. Sundown NP.

Genus *GLAPHYROMORPHUS*

11 Aust. spp.; 10 occur in Qld. Short limbs fail to meet when pressed to side of body (contacting on one sp.), all with 5 digits; movable lower eyelid lacking a transparent disc; no lobules on ear-opening; parietal scales in contact; frontoparietals divided; no pink ventral pigments; smooth shiny scales.

Moist habitats, including damp pockets within drier areas, along estn and nthn Qld. Also NT and WA. These secretive skinks seek humid shelter-sites under rocks, logs and leaf litter, and rarely forage in exposed areas, preferring sheltered, shaded environments. All Qld spp., except a pop. of *G. nigricaudis* from sthn CYP, are egglayers.

KEY TO *GLAPHYROMORPHUS*

a **b**

1 One lower labial scale contacts postmental scale **(a)**.......... 2
2 lower labial scales contact postmental scale **(b)** 6

2(1) Midbody scales 26 .... ***clandestinus*** (part)
Midbody scales 22 or fewer.. 3

3(2) A series of pale dorsolateral blotches.................. ***mjobergi***
No pale dorsolateral blotches 4

4(3) Upper flanks dark, clearly contrasting with paler back 5
Upper flanks not much darker than back, with no sharp contrast .... ***punctulatus***

5(4) Dark upper lateral zone fragmented by numerous pale spots and flecks .... ***pumilus***
Dark upper lateral zone solid and sharp-edged, with few or no pale spots and flecks .... ***crassicaudus***

6(1) Limbs widely-spaced, separated by noticeably more than a forelimb length when pressed to side of body 7
Limbs contacting or separated by no more than one forelimb length when pressed to side of body 8

7(6) Midbody scales 26... ***clandestinus*** (part)
Midbody scales 20–22 .... ***cracens***

8(6) A series of pale dorsolateral blotches..***fuscicaudis***
No pale dorsolateral blotches.. 9

9(8) Dark barred dorsal and lateral pattern extends back to groin; upper labials mainly dark ....
.... ***nyanchupinta***
Dark barred dorsal and lateral pattern restricted to forebody; upper labials mainly pale with dark sutures..
.... 10

10(9) 8 upper labial scales with 6th below centre of eye .... ***othelarrni***
7 upper labial scales with 5th below centre of eye .... ***nigricaudis***

Mount Elliot Mulch-skink

Glaphyromorphus clandestinus

SVL 72mm

Glaphyromorphus clandestinus. Mt Elliot.

Long body and tail with very short widely spaced limbs separated by noticeably more than one forelimb length when pressed to sides of body; midbody scales in 26 rows; one or 2 lower labial scales contacting postmental scale; dark longitudinal lines on flanks. Bronze to brown above with light to heavy dark speckling, barred lips and dark flecks on flanks aligned to form 7–8 dark stripes between ear and hindlimb.

Endemic. Recorded from moist humus under granite slabs at Alligator Ck Falls at an altitude of 425m on Mt Elliot in Bowling Green Bay NP in nthn BB. This sole known locality, essentially a WT outlier, is set in a mosaic of rainforest and eucalypt woodland.

Glaphyromorphus cracens. Millstream Falls.

Slender Mulch-skink
Glaphyromorphus cracens
SVL 58mm

Long body with very short, widely spaced limbs separated by more than several forelimb lengths when pressed to side of body; 2 lower labial scales contacting post-mental scale. Pale to dark brown above, with prominent, broad, black upper lateral stripe.

Endemic. Dry sclerophyll forests, woodlands, rock outcrops of EIU; estn limit probably marked by moister climate of WT.

Glaphyromorphus crassicaudus. Moa Is.

Northern Mulch-skink
Glaphyromorphus crassicaudus
SVL 55mm

Long body and tail; very short, widely spaced limbs separated by more than one forelimb length when pressed to side of body; one lower labial scale contacting postmental scale. Brown above, with dark paravertebral dashes sometimes forming complete lines or joined to form broad dark vertebral stripe, solid dark upper lateral stripe.

Wide variety of forests, woodlands, disturbed areas from Chillagoe in EIU through WT, CYP to Torres Strait and sthn NG. Separate ssp. occurs in ne. NT.

Brown-tailed Bar-lipped Skink
Glaphyromorphus fuscicaudis
SVL 90mm

Relatively well-developed limbs, separated by no more than one forelimb length when pressed to side of body; 2 lower labial scales contacting postmental scale; robust build;

Glaphyromorphus fuscicaudis. Lake Barrine.

pale dorsolateral blotches. Brown, darker on tail, with irregular dark bands on neck and anterior body, separated on sides of neck and shoulders by dorsolateral series of cream to pale yellow blotches. Ventral surfaces cream to yellow, most intense and sometimes merging to pink posteriorly.

Endemic. Rainforests, associated clearings in WT, from Paluma Ra. n. nearly to Cooktown.

S. Macdonald

Glaphyromorphus mjobergi. Lamb Range.

Atherton Tableland Mulch-skink

Glaphyromorphus mjobergi

SVL 90mm

Long body and tail, with very short, widely spaced limbs separated by at least 2 forelimb lengths when pressed to side of body; one lower labial scale contacting postmental scale; pale dorsolateral blotches. Brown above with dorsolateral series of cream to pale yellow blotches alternating with dark blotches from neck to shoulders or midbody.

Endemic. Montane rainforests above 650m between Ravenshoe area and Mt Carbine Tableland, WT. Shelters in and under moist rotting logs.

Black-tailed Bar-lipped Skink

Glaphyromorphus nigricaudis

SVL 90mm

Relatively well-developed limbs, separated by no more than one

Glaphyromorphus nigricaudis. Iron Range.

forelimb length when pressed to side of body; 2 lower labial scales contacting postmental scale; dark bands on anterior body; robust build. Brown with irregular, narrow dark bands across nape and forebody, fading and breaking at about midbody.

WT, estn and nthn CYP to sthn NG. Also ne. NT. Largely associated with damp pockets within woodlands, vine thickets, monsoon forests. Presumably egglaying over most of range, with a livebearing pop. on sthn CYP.

McIlwraith Bar-lipped Skink

Glaphyromorphus nyanchupinta

SVL 53mm

Relatively well-developed limbs separated by no more than one forelimb length when pressed to side of body; 2 lower labial scales contacting postmental scale; 7 upper labial scales with 5th below eye.

C. Hoskin

Glaphyromorphus nyanchupinta. McIlwraith Range.

Brown with dark wavy bars across dorsal and lateral surfaces, most prominent anteriorly, breaking up on flanks but extending back to hind limbs. Upper labial scales mainly dark, each with a pale central dot.

Endemic. Rainforest at McIlwraith Range, CYP. Recorded under logs at 530 metres altitude, but probably more widespread in uplands.

C. Hoskin

Glaphyromorphus othelarrni. Melville Range.

Cape Melville Bar-lipped Skink

Glaphyromorphus othelarrni

SVL 93mm

Well-developed limbs contact when pressed to sides of body; 2 lower labial scales contacting postmental scale; 8 upper labial scales with 6th below eye. Pale to dark brown above, with or without dark spots or wavy bars on anterior body. Flanks with dark, longitudinally aligned flecks or wavy vertical bars, strongest on sides of neck and breaking into flecks from forebody to groin and base of tail. Upper labial scales mainly pale with dark vertical bars along sutures.

Endemic. Isolated upland rainforest and lowlands atop the boulder fields of Melville Range, CYP. Recorded from 110–600 metres altitude. Associated with thick leaf litter among boulders.

Glaphyromorphus pumilus. Lake Wicheura.

Dwarf Mulch-skink

Glaphyromorphus pumilus

SVL 55mm

Long body and tail, with very short, widely spaced limbs separated by more than one forelimb length when pressed to side of body; one lower labial scale contacting postmental scale; pale flecks in upper flanks. Shades of brown to greyish brown with broken to continuous paravertebral lines of dark dashes, and dark, ragged-edged upper lateral zone fragmented by numerous pale flecks.

Endemic. Variety of habitats incl. woodlands, coastal dunes, rainforest edges, rock outcrops of estn CYP, from Isabella Falls to Torres Strait.

Fine-spotted Mulch-skink

Glaphyromorphus punctulatus

SVL 70mm

Long body and tail with very short, widely spaced limbs separated by at least several forelimb

Glaphyromorphus punctulatus. Warrawee Stn.

lengths when pressed to side of body; usually only one lower labial scale contacting postmental scale. Brownish grey to brown. Some pops prominently marked with simple dark flecks or spots, others patternless.

Endemic. Woodlands, vine thickets, rock outcrops from sthn WT, through EIU, BB, CQC to nthn SEQ.

Genus *GNYPETOSCINCUS* Monotypic genus.

Prickly Forest Skink

Gnypetoscincus queenslandiae

SVL 84mm

Strongly keeled dorsal and ventral scales with those on back and sides bead-like, and those on top of head rugose; well-developed limbs with 5 digits; parietal scales in contact; movable lower eyelid with no transparent disc; no lobules on large ear opening. Dark brown, sombrely marked with broken pale bands, often restricted to flanks.

Endemic. WT rainforests, from Cardwell Ra. n. to Rossville. Shelters in and under rotting logs, and can be extremely abundant. In some areas virtually all suitable sites occupied.

Gnypetoscincus queenslandiae. Wooroonooran NP.

Slow-moving, shuns direct sunlight, rarely encountered foraging. Bead-like scales readily disperse moisture across skin surface, possibly structured for that purpose. Livebearer; up to 6 young. Diet includes arthropods, gastropods, earthworms.

Genus *HARRISONIASCINCUS* Monotypic genus.

Beech Skink

Harrisoniascincus zia

SVL 55mm

Movable lower eyelid enclosing transparent disc; parietal scales in contact; frontoparietal scales divided; interparietal scale large; nasal scales widely separated; short well-developed limbs, each with 5 digits; ear opening present, without lobules. Brown with scattered dark and pale flecks, ill-defined, dark-edged pale dorsolateral line, dark upper flanks merging with paler mid to

Harrisoniascincus zia. Bithongabel, Lamington NP.

lower flanks, reddish brown flush on tail, brightest towards tip, bright yellow ventral surfaces.

Highlands of SEQ n. to Cunningham's Gap. Also ne. NSW. Restricted to rainforests and antarctic beech forests subject to low temperatures and high rainfall. Secretive, usually encountered sunning amongst leaf litter along edges of tracks and in clearings. Egglayer.

Forest Skinks
Genus *KARMA*

2 Aust. spp.; both occur in Qld. Well-developed limbs overlap when pressed to side of body, all with 5 digits; one lower labial scale contacts postmental scale; 3rd pair of chin shields separated by 5 longitudinal rows of smaller scales; scales along top of 4th toe in 2 rows with oblique sutures along almost entire length; movable lower eyelid lacking a transparent disc; no lobules on ear opening; parietal scales in contact; frontoparietals divided; smooth, shiny scales.

Rainforests of SEQ, incl. Border Ras where both spp. overlap. Also ne. NSW. These diurnal skinks are often seen beside walking tracks, perching on mossy logs and rocks or resting with head and forebody protruding from a hole or crevice. They tend to be confiding and approachable. Livebearers.

Murray's Skink
Karma murrayi
SVL 108mm

Last upper labial scale entire; flanks dusted with fine pale dots. Brown with fragmented dorsal pattern of dark flecks (occasionally forming ill-defined transverse bars), several large black lateral blotches between ear and forelimbs, fine dusting of pale bluish white dots, one or more per scale, on flanks. Ventral surfaces pale yellow between forelimbs and hindlimbs, with dark sutures to chin shields and oblique dark markings on chest.

Karma murrayi. Lamington NP.

Subtropical rainforests of SEQ, n. to Conondale Ra. Also ne. NSW. Lives in large rotting logs or rock crevices, and is often seen basking beside popular walking tracks.

Tryon's Skink
Karma tryoni
SVL 104mm

Rear upper labial scale divided into 2; flanks marked with scattered light and dark scales. Brown with dark, transverse, often broken dorsal bands continuous with row of dark dorsolateral blotches, several large black lateral blotches between ear and forelimbs, grey flanks enclosing scattered black and white scales. Ventral surfaces rich enamel yellow between forelimbs and hindlimbs, with obscure dark blotches on throat and very narrow dark sutures between chin scales.

SEQ. Restricted to closed, subtropical rainforests of Lamington Plateau, straddling Qld/NSW border. Commonly seen with head protruding from rock crevices or hollows in rotten logs and buttresses beside walking trails. It shares this habitat with closely related, very similar but much more widespread *K. murrayi*.

Karma tryoni. Lamington NP.

Sunskinks

Genus *LAMPROPHOLIS*

14 Aust spp; 12 occur in Qld. Four well-developed limbs, each with 5 digits; movable lower eyelid enclosing transparent disc; parietal scales in contact; frontoparietal scales fused to form single shield; interparietal scale distinct (fused on *L. adonis*); nasal scales widely separated; no lobules on ear opening.

Widespread in moist estn areas, n. to base of CYP. Also estn NSW to Tas and se. SA. Some spp. are restricted to rainforests and edges, others are amongst the most common lizards in parks and gardens. They are small, terrestrial and diurnal, basking on path edges and skittering amongst leaf litter when disturbed. All are egglayers, often depositing clutches of up to 7 eggs in communal caches that can number 100 or more.

KEY TO *LAMPROPHOLIS*

1	Interparietal scale fused to frontoparietals (**a**).........	***adonis***
	Interparietal scale free (**b**)	2
2(1)	Supraciliary scales 5 (**c**); usually fewer than 23 midbody scale rows..	***amicula***
	Supraciliary scales more than 5 (**d**); usually 23 or more midbody scale rows...	3
3(2)	Dark vertebral stripe present; usually 6 supraciliary scales.....	***guichenoti***
	No dark vertebral stripe; usually 7 supraciliary scales.........	4
4(3)	Limbs long, overlapping by much more than half a forelimb length when adpressed; dorsal pattern usually prominent, dominated by dark blotches and pale spots.......................................	***mirabilis***
	Limbs moderate, just failing to meet, or overlapping by less than half a forelimb length when adpressed; dorsal pattern weak, without conspicuous dark blotches and pale spots..	5
5(4)	Grey below with black pigment clustered under tail to form dark-edged pale blotches; high-altitude WT only	6
	White to yellow below; if grey, then dark pigment (if present) may form small spots or thin lines under tail but never clusters or dark-edged pale blotches	7

6(5)	Thornton Peak and uplands of Carbine Tableland	***robertsi***
	Bellenden Ker Ra. and highest parts of sthn Atherton Tablelands	***bellendenkerensis***
7(5)	Dark upper and pale lower flanks delineated, either by a pale midlateral stripe or a change of colour (usually abrupt)	8
	Dark upper flanks merge gradually with pale lower flanks; never a pale midlateral stripe	9
8(7)	Ventral surfaces usually yellow; Bunya Mtns only	***colossus***
	Ventral surfaces usually cream	***delicata***
9(7)	Flanks usually with some paler flecks; nthn Qld rainforests	10
	Flanks uniform; sthn Qld rainforests	***couperi***
10(9)	High elevations on Mt Elliot only	***elliotensis***
	WT and Townsville area	11
11(10)	Northern WT, nth of a line between Mareeba and Lake Barrine, along Gillies Range and Gordonvale, and se. to mouth of Russell R. incl. Malbon Thompson Ra	***coggeri***
	Sthn WT south of the above line to Townsville area	***similis***

R. Browne-Cooper

Lampropholis adonis. Eungella NP.

Lampropholis amicula. Beerwah.

Diamond-shielded Sunskink

Lampropholis adonis

SVL 50mm

Interparietal scale fused with frontoparietals to form one large shield; dark upper and pale lower flanks grade evenly with no abrupt demarcation. Brown above with contrasting grey flanks, colours separated by weak, narrow, pale dorsolateral line. On breeding males, throat becomes washed with blue, flanks and sides of tail are flushed with red, red reticulations develop beneath tail.

Endemic. Rainforest edges and clearings, lush creekside vegtn, in CQC, SEQ; isolated BB pop. in moist, shaded gorges of Carnarvon Ra.

Friendly Sunskink

Lampropholis amicula

SVL 35mm

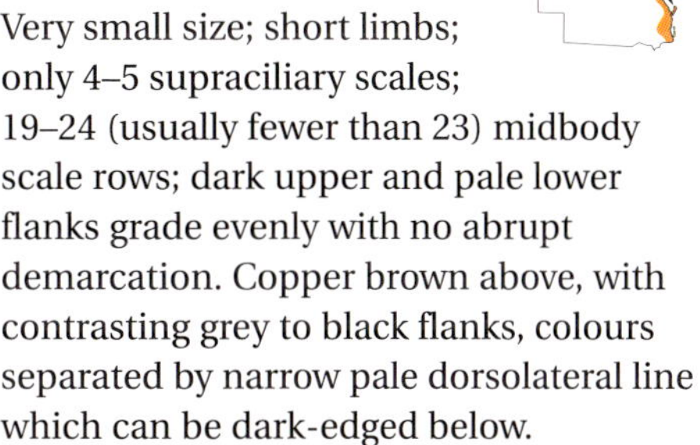

Very small size; short limbs; only 4–5 supraciliary scales; 19–24 (usually fewer than 23) midbody scale rows; dark upper and pale lower flanks grade evenly with no abrupt demarcation. Copper brown above, with contrasting grey to black flanks, colours separated by narrow pale dorsolateral line which can be dark-edged below.

SEQ, BB, inland to Bunya Mtns. Also ne. NSW. Dry sclerophyll forests, heathlands, vine thickets. Secretive, seldom straying far from leaf litter. Forages on and under mats of leaves, disappearing quickly if approached, much like *Menetia* and some small smooth-scaled *Carlia*.

A. Zimny

Lampropholis bellendenkerensis. Mt Bellenden Ker.

Lampropholis coggeri. Thornton Peak.

Mountain Sunskink
Lampropholis bellendenkerensis
SVL 47mm

Black ventral pigment incl. dark-edged pale spots under tail; 7 (occasionally 8) supraciliary scales. Brown above, sometimes with black dashes, delineated from dark brown upper flanks by narrow pale dorsolateral line. Upper and lower flanks merge or are clearly delineated; may have broken midlateral line of pale scales. Ventral surfaces grey with varying amount of black spots particularly on posterior body. A row of dark-edged pale blotches under tail. Distinguishable from *L. robertsi* only by distribution and genetics.

Endemic. High altitudes of WT above 900m on Bellenden Ker Ra. and upper elevations of sthn Atherton Tablelands. Associated with sunny edges and clearings.

Lampropholis colossus. Bunya Mtns.

Northern Wet Tropics Sunskink
Lampropholis coggeri
SVL 44mm

Short snout; 7 supraciliary scales; limbs narrowly separated or overlapping when adpressed; dark upper and pale lower flanks grade evenly with no abrupt demarcation. Brown to reddish brown above, usually with weak cream spots and black dashes. Flanks darker with pale spots and dark dashes, dorsal and lateral surfaces separated by weak, often ragged, narrow, pale dorsolateral stripe. Ventral surfaces spotted with black, tending to merge into reticulations under tail. Distinguishable from *L. similis* only by distribution and genetics.

Endemic. Nthn WT from Big Tableland s. to a line extending from Mareeba to Lake Barrine, along Gillies Range and Gordonvale, and se. to mouth of Russell R. incl. Malbon Thompson Ra. Rainforest clearings and edges; basks in sunny patches.

Bunya Mountains Sunskink
Lampropholis colossus
SVL 56mm

Very large size; 7 supraciliary scales; sharply delineated dark upper and pale lower flanks; yellow ventral pigment. Reddish brown to olive brown above, pale dorsolateral stripes and black laterodorsal stripes often broken into dashes, upper flanks dark brown to black, clearly defined from paler lower flanks by variably developed pale midlateral stripe, or abrupt colour change, ventral surface cream to bright yellow, speckled beneath tail with black, sometimes forming broken lines.

Endemic to uplands of Bunya Mtns, BB, inhabiting rainforests, bunya pine associations, tussock grass meadows. Status uncertain; possibly outlying pop. of *L. delicata.*

Lampropholis couperi. Mt Nebo.

Couper's Sunskink

Lampropholis couperi

SVL 49mm

Extensive black ventral pigment; 7 supraciliary scales; dark upper and pale lower flanks grade evenly with no abrupt demarcation. Brown above, usually with sparse dark dashes, upper flanks dark grey to black, contrasting dorsal and lateral colours separated by narrow pale dorsolateral line, ventral surfaces grey with black spots on chin and throat, black flecks on belly, black reticulations under tail.

Endemic. Rainforests on coastal ranges of SEQ, CQC. Basks in sunny patches, forest edges, clearings, margins of lush, rocky creek banks.

Lampropholis delicata. Kurwongbah.

Garden Skink

Lampropholis delicata

SVL 51mm

Extremely variable; 6–8 (usually 7) supraciliary scales; usually sharply delineated dark upper and pale lower flanks. Brown to olive brown above, often with dark and pale flecks, dashes and streaks. Pale dorsolateral line present, broken or absent. Dark upper flanks usually clearly defined from paler lower flanks by variably developed pale midlateral stripe, or an abrupt colour change, ventral surfaces cream with little or no dark pigment.

Broad range of relatively moist habitats, from rainforests to inner-city gardens, weedy vacant allotments in WT, EIU, BB, CQC, SEQ, NET. Also se. Aust. Extremely tolerant to human disturbance. Only Aust. lizard sp. to establish feral overseas populations, with established colonies in Hawaii, NZ, Lord Howe Is.

The abrupt colour difference between upper and lower flanks, cited as diagnostic in the key, not always discernible on some weakly marked individuals, leading to confusion with *L. couperi* and *L. coggeri.* These both have extensive dark ventral pigment, particularly under the tail, and predominate within their restricted rainforest habitats.

A. Zimny

Lampropholis elliotensis. Mt Elliot, Qld.

Mount Elliot Sunskink
Lampropholis elliotensis
SVL 40mm

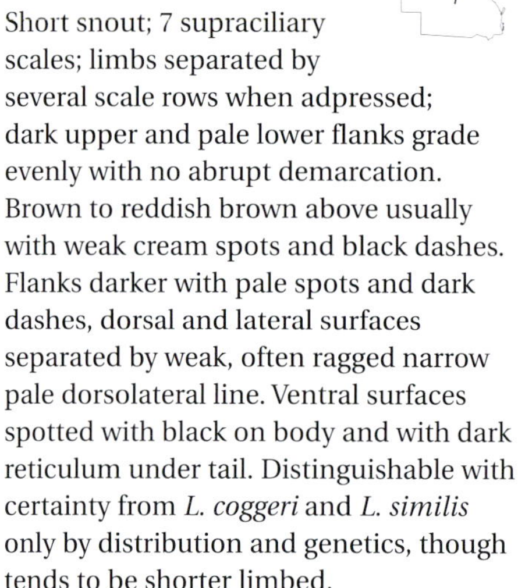

Short snout; 7 supraciliary scales; limbs separated by several scale rows when adpressed; dark upper and pale lower flanks grade evenly with no abrupt demarcation. Brown to reddish brown above usually with weak cream spots and black dashes. Flanks darker with pale spots and dark dashes, dorsal and lateral surfaces separated by weak, often ragged narrow pale dorsolateral line. Ventral surfaces spotted with black on body and with dark reticulum under tail. Distinguishable with certainty from *L. coggeri* and *L. similis* only by distribution and genetics, though tends to be shorter limbed.

Endemic. Mt Elliot in Bowling Green Bay NP, above 600m, sunning among leaf-litter on rocky areas.

Grass Skink
Lampropholis guichenoti
SVL 48mm

Dark vertebral stripe; 5–7 (usually 6) supraciliary scales; usually sharply delineated dark upper and pale lower flanks. Olive brown to grey with copper head, dark flecks and pale scales on back, ragged-edged (often indistinct) dark vertebral stripe from nape to base

Lampropholis guichenoti. Cooloola NP.

of tail. Dark upper flanks separated from back by narrow pale dorsolateral line, and from pale lower flanks by an abrupt colour change or weak to very conspicuous pale midlateral stripe, ventral surfaces cream with little or no dark pigment.

Largely associated with moist heaths in NET, coastal SEQ. Widespread in se. Aust.

Lampropholis mirabilis. Mt Elliot.

Saxicoline Sunskink
Lampropholis mirabilis
SVL 50mm

Long limbs; dorsally depressed body; 5–8 (usually 7) supraciliary scales; usually sharply delineated dark upper flanks and pale lower flanks. Olive grey with copper head and prominent dark spots or blotches mixed with scattered pale flecks or spots over body, limbs and tail. Narrow, pale dorsolateral stripe and broken midlateral stripe sometimes present.

Endemic to Townsville district, far nthn BB. Rock-inhabiting; along rocky creek beds in rainforest and hoop pine on Magnetic Is., Cape Cleveland, Mt Elliot. Forages on boulders and rock faces, leaping with ease between them. Long limbs and flattened build are common features of rock-inhabiting skinks.

Lampropholis robertsi. Mt. Lewis. *S. Zozaya*

Roberts' Sunskink

Lampropholis robertsi

SVL 51mm

Black ventral pigment incl dark-edged pale spots under tail; 7 (occasionally 8) supraciliary scales. Brown above, sometimes with black dashes, delineated from dark brown upper flanks by narrow pale dorsolateral line. May have broken midlateral line of pale scales. Upper and lower flanks merge or are clearly delineated. Ventral surfaces grey with varying amount of black spots particularly on posterior body. A row of dark-edged pale blotches under tail. Distinguishable from *L. bellendenkerensis* only by distribution and genetics.

Endemic. WT above about 900m on Thornton Peak and Carbine Tablelands. Basks in sunny rainforest clearings, rocky areas and edges and montane heaths.

Lampropholis similis. Malanda Falls.

Southern Wet Tropics Sunskink

Lampropholis similis

SVL 42mm

Short snout; 7 supraciliary scales; limbs overlap or narrowly separated when adpressed; dark upper and pale lower flanks grade evenly with no abrupt demarcation. Brown to reddish brown with 4–6 lines of black dashes, sometimes obscure to absent in sthn pops. Flanks darker with pale spots and dark dashes, dorsal and lateral surfaces separated by weak, often ragged narrow pale dorsolateral line. Ventral surfaces spotted with black on body and with dark reticulum under tail. Distinguishable with certainty from *L. coggeri* and *L. elliotensis* only by distribution and genetics, though tends to be longer limbed than *L. elliotensis*.

Endemic. Sthn WT from Lake Barrine and Gillies R. area s. to Hervey Ra. and The Pinnacles near Townsville, incl. Palm and Hinchinbrook Is. Basks in rainforest clearings and edges.

Genus *LERISTA*

97 Aust. spp.; 25 occur in Qld. Limbs very small and widely spaced, or forelimbs absent, or all limbs absent; digits 4, 3, 2 or one, with fingers equal to or fewer than toes; lower eyelid movable and enclosing transparent disc, or fused to form fixed spectacle; ear openings present but minute; parietal scales in contact. Pattern is usually a simple combination of longitudinal lines of dark dorsal dashes, a dark upper lateral stripe and sometimes a reddish flush on tail.

Widespread in dry areas such as deserts, brigalow communities, and the woodlands and ranges of ne. interior. Also dry areas of all mainland states. Egglayers.

Lerista is often cited as a classic example of evolution in process, with progressive limb reduction and digit loss, coupled with modifications to snout-shape, facilitating different degrees of burrowing. Those with shovel-shaped snouts, no forelimbs and 2-toed hindlimbs swim through soft upper sand layers on desert dunes. Others with 4 moderately well-developed limbs and 3–4 fingers and toes inhabit leaf litter and soft soil under rocks and logs. *L. ameles* lacks limbs completely.

A *Lerista labialis*. Ballera.

B *Lerista storri*. Springfield Stn.

C *Lerista timida*. Augathella.

Variation in snout-shapes of *Lerista* species. *Lerista labialis* (**A**), a sandswimmer in loose exposed sand of dunes and plains has an acutely shovel-shaped snout. *Lerista storri* (**B**), with a covering of thickened scales to protect the snout, moves through loose sand under logs and leaf litter. *Lerista timida* (**C**) is active on the surface and through upper leaf litter profiles. Its rounded snout lacks specialised modifications.

KEY TO *LERISTA*

1	Four limbs clearly discernible	2
	Fewer than 4 limbs, or if 4, then forelimb reduced to a minute nubbin in a dimple	11
2(1)	Limbs each with 4 digits	3
	Limbs each with fewer than 4 digits	7
3(2)	Lower eyelid moveable	***chordae***
	Lower eyelid fused	4
4(3)	Reddish back; sw. deserts	***aericeps***
	Brown, grey or olive; ne. to nw	5
5(4)	Dark upper lateral stripe with sharp upper edge but diffuse lower edge	***orientalis***
	Dark upper lateral stripe with sharp upper and lower edges	6
6(5)	Lower lateral surfaces brownish white with darker dots or edges on scales; ne. interior	***zonulata***
	Lower lateral surfaces white and patternless; white sands on se. coastal CYP	***ingrami***

7(2)	Limbs each with 3 digits	8
	At least some limbs with fewer than 3 digits	9
8(7)	Lower eyelid movable	***fragilis***
	Lower eyelid fused	***timida***
9(7)	2 fingers on forelimb	10
	One finger on forelimb	***punctatovittata***
10(9)	3 toes on hindlimb	***desertorum***
	2 toes on hindlimb	***emmotti***
11(1)	At least some limbs present	12
	All limbs absent	***ameles***
12(11)	Hindlimb with 2 toes	13
	Hindlimb with one toe or represented by a stump	16
13(12)	Interparietal scale distinct **(a)**	14
	Interparietal and frontoparietals fused to form one shield **(b)**	15
14(13)	Third upper labial scale below eye	***wilkinsi***
	Fourth upper labial scale below eye	***allanae (part)***
15(13)	Two supraocular scales in contact with frontal **(c)**	***bipes***
	One supraocular scale in contact with frontal **(d)**	***labialis***
16(12)	Fourth upper labial scale below eye	17
	Third upper labial scale below eye	19
17(16)	Hindlimb stylar; CYP only	***anyara***
	Hindlimb clawed; BB, EIU and MGD only	18
18(17)	Postmental scale usually contacts 2 lower labial scales **(e)**; no forelimb; dorsal and lateral rows of dark spots; all ventral scales dark-edged	***allanae (part)***
	Postmental scale contacts one lower labial scale **(f)**; forelimb represented by a nubbin; continuous dark dorsal and lateral lines; dark flecks present or absent on ventral scales	***colliveri***
19(16)	Midbody scales in 16 rows	***rochfordensis***
	Midbody scales in 18 or more rows	20
20(19)	Dark lateral stripe present	21
	Dark lateral stripe absent	22
21(20)	Ventral surface darkly pigmented; Blackbraes NP	***vanderduysi***
	Ventral surface immaculate; Mt Cooper Stn area	***vittata***
22(20)	Two loreal scales	***hobsoni***
	One loreal scale	23
23(22)	Four supraciliary scales	24
	Two supraciliary scales	***parameles***
24(23)	Dorsal surface with dark dashes forming thin lines	25
	Dorsal surface mainly without pattern; Mt Surprise area	***storri***
25(24)	Supraciliary scales in a continuous line	***cinerea***
	Fourth supraciliary scale separated from other supraciliaries	***alia***

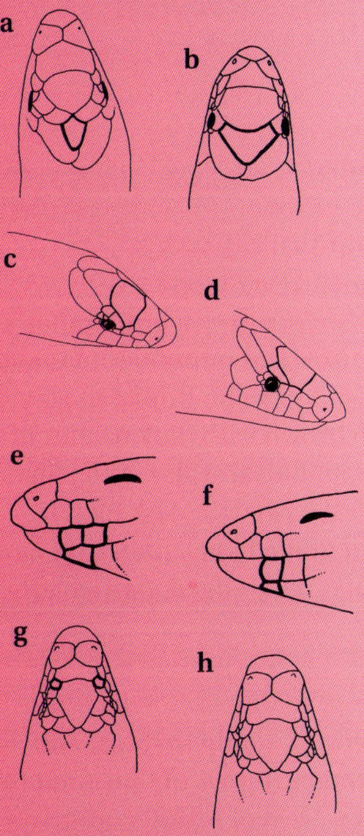

M. Sanders

Lerista aericeps. Windorah.

Yellow-tailed Slider

Lerista aericeps

SVL 52mm

Four well-developed limbs, each with 4 digits; fused lower eyelid. Reddish brown with 4 dorsal rows of dark spots, dark upper lateral stripe reduced on body to series of dark spots, often a bright yellow to copper flush on tail.

Sandy spinifex deserts in CHC. Also arid zones of NT, NSW, SA. Associated with soft sand on dune-crests and slopes. Taxonomic status of this and closely allied wstn sp., *L. xanthura*, remains uncertain.

Bulleringa Fine-lined Slider

Lerista alia

SVL 73mm

No trace of forelimb; hindlimb minute with one clawed toe or stylar; no prefrontal scales; 4 supraciliary scales; anterior chin shields separated by a scale; movable lower eyelid. Brown to silvery grey with prominent to obscure fine dark broken lines along body, more prominent on tail.

Lerista alia. Van Lees Stn.

Endemic. EIU, in loose soil under logs in woodlands to open woodlands from Bullaringa NP to Talaroo area.

Lerista allanae. Between Capella and Clermont.

Retro Slider

Lerista allanae

SVL 88mm

No forelimbs (usually only a depression and very rarely a nubbin); only one toe on hindlimb n. of Capella, 2 toes (2nd minute) s. of Capella; moveable lower eyelid. Brown to greyish or blackish brown with dark dashes or dark edges on dorsal scales, and paler lateral scales, each with dark blotch creating chequered appearance on flanks. Ventral scales dark-edged, particularly under tail.

Endemic. Currently known only from small areas of friable soils in vicinity of Logan Downs and Retro Stns, and road side verges from s. of Clermont to s. of Capella. Most of region comprises heavy dark soils that are probably unsuitable. Once feared extinct because region has undergone such extensive modification for agricultural and resource development. Existing pops are probably small and isolated.

Lerista ameles. Mt Surprise area.

Limbless Fine-lined Slider

Lerista ameles

SVL 58mm

No limbs, forelimbs completely absent and hindlimbs reduced to mere depressions; movable lower eyelid. Silvery grey to brown with narrow dark longitudinal lines from nape to tail.

Endemic to Mt Surprise area, EIU. Recorded in soft soil under rocks in open grassy woodlands with low weathered granite outcrops.

Lerista anyara. Kimba Plateau.

Olkola Slider

Lerista anyara

SVL 85mm

No forelimbs; hindlimb stylar; prefrontal scales present; 2 supraocular scales; 4 supraciliary scales in continuous row. Yellowish brown with dark dash in each scale aligning to form narrow ragged lines, most prominent dorsally, becoming broken and diffuse laterally, and forming broader and more continuous stripes on tail.

Endemic. CYP, in open woodland on sandy soil on Kimba Plateau in Olkola NP.

Lerista bipes. Barkly Stn area, NT.

Western Two-toed Slider

Lerista bipes

SVL 62mm

No trace of forelimbs; 2 toes on hindlimb; movable lower eyelid; flat, protrusive snout; 5 upper labial scales; frontal scale in contact with 2 supraocular scales; midbody scales in 18 rows. Yellowish brown to reddish brown with 2 dorsal lines of dark dashes and broad, prominent, dark upper lateral stripe.

Dunes and other areas of soft sand in spinifex deserts of CHC. Also arid zones of NT, SA, WA. Shelters under leaf litter; forages just below surface of exposed sand, leaving meandering trails.

Lyre-patterned Slider

Lerista chordae

SVL 45mm

Four well-developed limbs, each with 4 digits; moveable

E. Vanderduys

Lerista chordae. Woura Park Stn.

lower eyelid. Greyish brown to bronze with 4 lines of dashes from nape to base of tail and dark brown upper lateral stripe from snout to base of tail, breaking into flecks on tail.

Endemic. DEU between Torrens Ck and Aramac, on red and yellow sandy soils supporting woodlands, spinifex and other grasses.

Lerista cinerea. Warrawee Stn.

Vine-thicket Fine-lined Slider

Lerista cinerea

SVL 66mm

No trace of forelimbs; one toe on greatly reduced hindlimb; movable lower eyelid; prefrontal scales present. Silvery grey with dark dashes on each dorsal scale aligned to form narrow lines. Tail of juv. flushed with red, fading to yellow on adult.

Endemic to EIU. Known from acacia forest, eucalypt woodland, semi-evergreen vine thicket on dark clay loams, heavy red loams, yellow sandy soils and stony soils. Inhabits soft soils under rocks, logs, leaf litter.

G. Shea

Lerista colliveri. Red Falls.

Colliver's Slider

Lerista colliveri

SVL 84mm

Virtually no forelimbs (reduced to minute nubbin within a dimple); only one toe on hindlimb; movable lower eyelid. Brown to greyish brown with simple dark streak on each dorsal and lateral scale, aligning to form series of narrow lines which break into dark-edged scales on lower flanks. Ventral surfaces patternless or dark-flecked, with weak to strong longitudinal lines of dark flecks under tail.

Endemic. EIU, adj. MGD. Open woodlands and rock outcrops between Harvey Ra. and Hughenden area.

Great Desert Slider

Lerista desertorum

SVL 93mm

Four very short limbs with 2 digits on forelimb and 3 on hindlimb; moveable lower eyelid. Reddish brown to greyish brown with paravertebral and sometimes dorsal series of small dark spots from nape to base

E. Vanderduys

Lerista desertorum. Craven's Peak.

of tail, breaking on tail, and broad dark upper lateral stripe from snout to hips.

Recorded from acacia shrubland on stony plateau at Craven's Peak in CHC. Also sandy arid areas of NT, SA and WA.

Lerista emmotti. Noonbah Stn.

Noonbah Robust Slider

Lerista emmotti

SVL 103mm

Four very short limbs, each with 2 digits; movable lower eyelid. Fawn to brown, often with a dark streak on each scale aligning to form longitudinal rows of dark spots.

CHC, ML, MGD, n. to about Muttaburra and se. to Currawinya NP. Also nw. NSW, adj. SA. Very similar to *L. punctatovittata* but with extra finger. *L. emmotti* tends to occur further w. but in some areas (e g. Currawinya) both occur. Shelters under leaf litter at bases of trees and shrubs.

Lerista fragilis. Denham Park Stn.

Eastern Mulch Slider

Lerista fragilis

SVL 60mm

Four well-developed limbs, each with 3 digits; movable lower eyelid. Olive brown to silvery brown with 4–6 narrow lines of dark dorsal dashes, narrow dark upper lateral stripe and red flush on tail of juv.

Endemic. Widespread in dry sclerophyll forests, woodlands, shrublands of SEQ, BB, ML, MGD, DEU, EIU, NWH. Shelters in soil under rocks, logs, leaf litter.

Lerista hobsoni. Myola Stn.

Hobson's Fine-lined Slider

Lerista hobsoni

SVL 68mm

No trace of forelimbs; one minute toe on hindlimb; 2 loreal scales; prefrontal scale present; lower eyelid movable. Silvery grey with 6 narrow lines of dark dashes.

Endemic. Open forests in Upper Burdekin R. Drainage Basin in nthn BB.

Lerista ingrami. Cape Flattery.

McIvor River Slider
Lerista ingrami
SVL 36mm

Four well-developed limbs, each with 4 digits; fused lower eyelid; usually 18 (rarely 20) midbody scale rows. Very pale greyish brown to pinkish brown with black dot in centre of each dorsal scale forming 4 longitudinal rows, prominent dark brown upper lateral stripe with sharp, straight upper and lower edges, completely patternless white lower flanks and belly, pale to bright orange tail.

Endemic. Known only from white coastal sand, incl. first coastal dunes, near mouth of McIvor R. on se. CYP. Recorded in loose soft sand under coconuts and other debris.

Lerista labialis. Nappa Merrie Stn.

Eastern Two-toed Slider
Lerista labialis
SVL 60mm

No trace of forelimbs; 2 toes on hindlimb; movable lower eyelid; flat, protrusive snout; 5–6 upper labial scales; frontal scale in contact with only one supraocular; midbody scales in 18–20 rows. Yellowish brown to reddish brown with 2 dorsal lines of dark dashes and broad, prominent, dark upper lateral stripe.

Sandy deserts of CHC. Also sandy arid zones of all mainland States except Vic. Like similar *L. bipes,* a consummate sand-swimmer, sheltering in soft upper layers under debris, foraging in exposed sites such as dune-slopes and crests, leaving distinctive meandering trails.

S. Macdonald

Lerista orientalis. Doomagee area.

North-eastern Slider
Lerista orientalis
SVL 49mm

4 well-developed limbs, each with 4 digits; fused lower eyelid. Pale to dark olive grey to reddish brown, often becoming redder on tail, with dark dorsal dots forming 4 longitudinal rows, broad dark upper lateral stripe with sharp upper edges and diffuse, ragged lower edges.

Known in Qld from Doomagee and Mornington Is., GUP. Also nthn NT, adj. WA. Shrublands, woodlands on heavy, stony soils.

Chillagoe Fine-lined Slider
Lerista parameles
SVL 71mm

No trace of forelimb; hindlimb minute with one clawed

Lerista parameles. Almaden area.

digit; prefrontal scales absent; usually 2 supraciliary scales; anterior chin shields in broad contact; lower eyelid movable. Brown to silvery grey with dark lines, broken to continuous, along body becoming more prominent on tail.

Endemic. Sandy soils with woodlands to open woodlands in vicinity of Chillagoe in EIU.

Lerista punctatovittata. Dunmore SF.

Eastern Robust Slider

Lerista punctatovittata

SVL 100mm

Four very short limbs, with one finger, 2 toes; movable lower eyelid. Shades of brown with dark sutures on head shields, dark bars on lips and usually dark dashes on each scale aligning to form longitudinal rows.

Woodlands, shrublands in BB, ML, CHC, DEU, MGD. Also semi-arid to arid parts of wstn NSW, nw. Vic, estn SA.

Dwells in soft soil under logs and litter, mainly on sandy soils; also on heavy soils with brigalow, utilising thick mat of leaf litter around each tree.

Lerista rochfordensis. Amity Stn.

Rochford Slider

Lerista rochfordensis

SVL 82mm

No trace of forelimbs; one toe on greatly reduced hindlimb; moveable lower eyelid; striped pattern. Grey to beige with 8–10 narrow dark longitudinal lines, each running through the centre of a scale row.

Endemic. Isolated pops in vine thickets, comprising Lancewood (*Acacia shirleyi*) and Bendee (*A.catenulata*) on loose coarse sand. Each site covers only several hundred square m. and is separated by about unsuitable habitat, at Rochford Scrub and Boori Stn, se. of Charters Towers in nthn BB. Recorded in loose sand under logs.

Mount Surprise Fine-lined Slider

Lerista storri

SVL 66mm

No trace of forelimb; hindlimb stylar; prefrontal scales absent; 4 supraciliary scales; anterior

Lerista storri. Springfield Stn.

chin shields separated by a scale; lower eyelid movable. Brown to silvery grey with little or no dorsal pattern on body. Flanks with longitudinal lines and tail has prominent dorsal and lateral lines.

Endemic. Woodlands to open woodlands on Springfield Stn in Mt Surprise area in EIU.

Lerista timida. Girraween NP.

Dwarf Three-toed Slider

Lerista timida

SVL 50mm

Four well-developed limbs, each with 3 digits; fused lower eyelid. Olive brown to silvery brown with 4 narrow lines of dark dorsal dashes, narrow dark upper lateral stripe and yellow to red flush on tail of juv.

Widespread in dry habitats in NET, BB, ML, DEU, MGD, CHC, possible isolated pop. in NWH. Also dry to arid areas of all mainland States. Lives in soft soil under rocks, logs, leaf litter.

Lerista vanderduysi. Blackbraes NP.

Leaden-bellied Fine-lined Slider

Lerista vanderduysi

SVL 67mm

No trace of forelimb; hindlimb minute with 1 digit; prefrontal scales absent; lower eyelid movable. Rich glossy brown with 4 dorsal series of streaks and a broad, prominent dark lateral stripe from behind eye onto tail. Ventral surface dark, suffused with brown to grey.

Endemic. Vine thickets in Blackbraes NP and Gilberton Stn area, EIU.

Lerista vittata. Mt Cooper Stn.

Side-striped Fine-lined Slider

Lerista vittata

SVL 63mm

No trace of forelimbs; one toe on greatly reduced hindlimb; movable lower eyelid; prefrontal scales absent. Silvery grey to silvery brown with fine dashes on each dorsal scale forming narrow, dark broken lines along body and tail, and prominent dark brown lateral stripe. Juv. and subadult have orange flush on tail.

Endemic. Recorded in a small area of vine thickets on soft sandy soil at Mt Cooper Stn near Charters Towers.

Lerista wilkinsi. White Mountain NP.

Wilkin's Slider
Lerista wilkinsi
SVL 75mm

No trace of forelimbs; 2 toes on greatly reduced hindlimb; movable lower eyelid. Silvery grey to silvery brown with dark dashes on each scale forming narrow longitudinal lines along body and tail. Tail of juv. flushed with red.

Endemic. Nthn DEU, in rocky areas of Warrigal Ra. and White Mtns. Inhabits soft sand and leaf litter accumulated in rocky depressions, around bases of trees.

Lerista zonulata. Hughenden area.

Wide-striped Four-toed Slider
Lerista zonulata
SVL 50mm

Four well-developed limbs, each with 4 digits; fused lower eyelid. Pale to dark olive brown, redder on tail, with 4–6 dorsal rows of small dark dots, broad dark upper lateral stripe with sharp, straight, upper and lower edges. Scales on lower flanks brownish white, dark-edged.

Endemic. Dry, stony, hilly areas of far nthn BB, EIU, GUP, approaching wstn edge of WT at Windsor Tableland. Lives in loose soil under rocks, logs, leaf litter.

Four-fingered Rock Skinks
Genus *LIBURNASCINCUS*

4 Aust. sp.; all occur in Qld. Long, well developed limbs, with 4 fingers and 5 toes; moveable lower eyelid enclosing a transparent disc; ear opening present and conspicuous with lobules; parietal scales in contact; frontoparietals fused to form a single shield; dorsal scales keeled, with keels sometimes formed by rows of small raised points.

Restricted to rocky habitats in ne. Qld. Swift, extremely agile, alert skinks, and skilled climbers performing well-coordinated leaps between boulders. Highly visually cued, employing tail-waving as signals. Sun-lovers. Egglayers.

KEY TO *LIBURNASCINCUS*

1 Keels on dorsal scales usually comprise a row of small points; paravertebral scales 59 or more 2
No raised small points on keels on dorsal scales; paravertebral scales 56 or fewer 3

2(1) Ear aperture with small rounded lobules on margins; juv. and all except large adults with prominent wavy lines and spots .... ***coensis***
Ear aperture surrounded by long, sharp lobules; adults and juv. with little or no pattern .... ***scirtetis***

3(1) An even gradation from large inner to smaller outer plantar scales on foot..................... ***mundivensis***
Inner row of plantar scales about twice as large as outer planter scales, with an abrupt change in size .... ***artemis***

A. Zimny

Liburnascincus artemis. Bamboo Range.

Liburnascincus coensis. McIlwraith Ra.

Bamboo Range Rock Skink
Liburnascincus artemis
SVL 58mm

2 weak to moderate keels on dorsal scales (some 3-keeled scales usually present), each hexagonal with angular hind edge; moderate-sized disc in lower eyelid; ear-opening round, surrounded by conspicuous pointed lobules, largest anteriorly; 34–36 midbody scales (mean 35); inner row of plantar scales large and block-like, abruptly larger than remaining densely packed plantar scales on foot. Grey, copper on head, with paired paravertebral series of dark and pale spots. Vertebral region patternless.

Endemic. Low rocky hills on Bamboo Ra. in CYP. but probably more continuous along similar habitat.

Coen Rock Skink
Liburnascincus coensis
SVL 68mm

Longitudinal rows of small tubercles on dorsal scales, each with curved hind edges; small disc in lower eyelid; vertically elongate ear opening with small round lobules around margins; large size; long limbs. Very dark brown with prominent gold or silver vertebral, dorsolateral and midlateral stripes or rows of blotches.

Endemic. Perches conspicuously on boulders in vine thickets along creeks in McIlwraith and Table Ra., between Coen and Pascoe R., ne. CYP.

Outcrop Rock Skink

Liburnascincus mundivensis

SVL 56mm

2 or 3 moderately strong keels on dorsal scales, each hexagonal or with curved hind edge; small disc in lower eyelid; round ear opening surrounded by pointed lobules; 34–42 midbody scales (mean 38); large inner plantar scales grade evenly into smaller outer plantar scales on foot. Olive brown with numerous large dark blotches over all but pale vertebral zone. Pale flecks often present, arranged to form vague vertebral, dorsolateral and midlateral stripes, particularly on juv.

Endemic. Occupies boulders and rock faces in dry, well-drained areas of BB, EIU, wstn edge of CQC and CYP between about Gladstone and inland from Cooktown.

Liburnascincus mundivensis. Amity Stn.

Liburnascincus scirtetis. Black Mountain.

Black Mountain Skink

Liburnascincus scirtetis

SVL 64mm

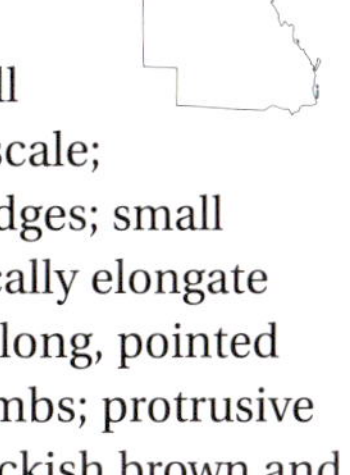

Longitudinal rows of small tubercles on each dorsal scale; scales with curved hind edges; small disc in lower eyelid; vertically elongate ear opening margined by long, pointed lobules; large size; long limbs; protrusive eyes; upturned snout. Blackish brown and patternless, or with paler greenish scales tending to form vague vertebral and dorsolateral stripes, particularly on juv.

Endemic to piled black granite boulders of Black Mtn in far nthn WT. Abundant, conspicuous in early morning, late afternoon, basking, tail-lashing and head-bobbing on exposed rock surfaces. Inquisitive, regarding humans with keen but cautious eye.

Genus *LIOPHOLIS*

11 Aust. spp.; 3 occur on Qld. Widely separated parietal scales; 4 well-developed limbs each with 5 digits; 4th toe much longer than 3rd; moveable lower eyelid with no transparent disc; eyes relatively large with distinctive cream margins to lids; ear opening present with obvious anterior lobules; smooth scales, moderate size.

Sthn Qld, in temperate to arid habitats. Shelters range from burrows excavated under shrubs on dune slopes to modified cavities in soil under rocks and logs. Basking often occurs near burrow entrance. Omnivorous live-bearers, often dwelling in colonies.

KEY TO *LIOPHOLIS*

1 Head very deep with short, blunt snout; sw. arid zone.. ***inornata***
Head and snout not deep and blunt; woodlands or outcrops of se. or estn interior............................ 2

2 Usually a pale streak through upper lips; prominent dark-edged pale spots on flanks; dorsal pattern, if present, prominent and including two longitudinal rows of pale spots................................. ***whitii***
No pale streak through upper lips; lateral pattern usually weak and pale spots not dark-edged; dorsal pattern absent or reduced to narrow dark lines .... ***modesta***

Liopholis inornata. Ethabuka.

Liopholis modesta. Lake Broadwater.

Desert Skink
Liopholis inornata
SVL 84mm

Short, blunt head; moderately small size. Reddish brown with black spots scattered or arranged in longitudinal rows on back, in vertical rows on flanks, and forming dark bars on sides of tail.

Sandy deserts of CHC, ML, e. to Currawinya NP. Also arid zones of all mainland States. Mainly assoc. with dunes and swales, excavating burrows in sloping sand, usually at base of bush or spinifex hummock, normally with obvious entrance and one or more concealed exits. Crepuscular to nocturnal.

Eastern Ranges Rock-skink
Liopholis modesta
SVL 112mm

Weak pattern with no pale streak on upper lips. Pale brown to greyish brown with pale ear lobules, and few spots on upper lip. Specimens from interior incl. Yuleba, Roma and Chesterton Ra. have rich red dorsal and ventral flush. Pattern absent or comprising narrow dark dorsal lines and widely spaced pale lateral spots sometimes restricted to anterior flanks.

Dry woodlands, rock outcrops of SEQ, BB, NET, e. to Lockyer Valley. Also ne. and central NSW. Excavates shallow burrows and depressions under rocks and logs.

White's Skink

Liopholis whitii

SVL 113mm

Moderately small size; dark-edged ocelli on flanks; pale streak on upper lip. Patterned forms have broad tan vertebral stripe, broad darker brown to black laterodorsal stripes enclosing row of pale spots, pale dorsolateral stripes, grey flanks, and lateral series of large, dark-edged pale spots incl. cluster above forelimb. Plain-backed forms retain lateral pattern but have unmarked reddish brown backs.

NET only in Qld. Also estn NSW, Vic, SA, Tas. Excavates burrows or shallow depressions under rocks.

Liopholis whitii. Plain form. Girraween NP.

Liopholis whitii. Patterned form. Girraween NP.

Litter-skinks
Genus *LYGISAURUS*

12 Aust spp.; all occur in Qld. Well-developed limbs with 4 fingers and 5 toes; moveable lower eyelid enclosing a transparent disc (fixed spectacle on one sp.) ranging from large (much more than half size of eyelid) to small (about half size of eyelid or less); ear opening conspicuous, usually with lobules; parietal scales in contact; frontoparietal scale fused to form one shield; dorsal scales smooth (with rows of small points on 2 spp.).

Estn Qld with centre of diversity in ne. Also estn NSW and ne. NT. Very small, swift, secretive sun-loving skinks that forage among leaf litter and seldom venture into open spaces. Breeding males develop reddish flushes on hips, tail and sometimes throat. Egglayers, 2 per clutch.

KEY TO *LYGISAURUS*

1	Lower eyelid fused to form a spectacle	***foliorum***
	Lower eyelid moveable	2
2(1)	Dorsal scales with 4–5 weak carinations comprising series of small points	3
	Dorsal scales smooth	4
3(2)	Pale vertebral stripe	***parhassius***
	No pale vertebral stripe	***rimula***
4(2)	6 upper labials	5
	7 upper labials	6

a **b**

5(4) Ear opening round; edged by flat or blunt lobules **(a)** .... ***macfarlani***
Ear-opening horizontal; edged by sharp lobules **(b)** ***aeratus***

6(4) Ear opening edged by sharp lobules **(b)**........ 7
Ear opening edged by flat or blunt lobules **(a)**........ 10

7(6) Palpebral disc large, occupying more than half of lower eyelid 8
Palpebral disc small, occupying less than half of lower eyelid........ 9

8(7) Hindlimbs and tail flushed with reddish orange; no discernible spotting under hindlimbs and tail .... ***malleolus***
No reddish orange flush on hindlimbs and tail; spotting visible under hindlimbs and tail...... ***abscondita***

9(7) Pale dorsolateral stripes on juveniles, females and non-breeding males; nthn CYP .... ***sesbrauna***
No pale dorsolateral stripes; sthn CYP .... ***laevis***

10(6) Supraciliary scales 7........ ***tanneri***
Supraciliary scales 6........ 11

11(10) Midbody scales 23–24 .... ***zuma***
Midbody scales 27 or more........ ***rococo***

C. Dollery

Lygisaurus abscondita. Donkey Springs.

Lygisaurus aeratus. Earlville.

Secretive Litter-skink

Lygisaurus abscondita

SVL 30mm

Large disc in lower eyelid; round to slightly horizontally elongate ear opening surrounded by numerous sharp lobules; 7 upper labial scales. Dark olive brown, paler on hips and tail, with dark bars on lips and faint dark speckling under hindlimbs and tail.

Endemic. Open woodlands and sandstone areas in vicinity of Bulleringa NP, incl. Donkey Springs, in EIU.

Large-disced Litter-skink

Lygisaurus aeratus

SVL 39mm

Large disc in lower eyelid; usually horizontally elongate ear opening surrounded by numerous sharp lobules; usually 6 upper labial scales. Olive brown above, copper on head, with dark flanks often speckled with white. On breeding males, throat, hindlimbs and tail flushed with red.

Endemic. Secretive inhabitant of leaf litter in woodlands, grasslands and rock outcrops in WT, EIU, CYP between Ingham and Rokeby Stn area.

Lygisaurus foliorum. Kurwongbah.

Tree-base Litter-skink
Lygisaurus foliorum
SVL 39mm

Lower eyelid fused to form a fixed spectacle; ear opening horizontally elongate with one or more sharp to low flat lobules; usually 7 upper labial scales. Brown to greyish brown with little pattern. On breeding males, throat, hindlimbs, tail flushed with pinkish orange.

Dry sclerophyll forests and woodlands in NET, SEQ, BB, CQC, DEU, EIU, WT. Also estn NSW. Secretive, actively sunning and foraging amongst leaf litter, seldom venturing far from cover.

Lygisaurus laevis. Kuranda.

Rainforest Edge Litter-skink
Lygisaurus laevis
SVL 37mm

Small disc in lower eyelid; horizontally-elongate ear opening surrounded by numerous sharp lobules of approx. equal size; 7 upper labial scales. Reddish brown above (shining iridescent green/purple from some angles), sometimes with narrow dark lines, and darker flanks, usually enclosing pale flecks. On breeding males, throat and tail flushed with red.

Endemic. Sunlit edges of tropical rainforests in WT, between Cooktown and Cardwell Ra.

Lygisaurus macfarlani. Breeding male. Badu Is.

Translucent Litter-skink
Lygisaurus macfarlani
SVL 37mm

Small disc in lower eyelid; round ear opening with a large blunt anterior lobule and several small flat lobules around margin; usually 6 upper labial scales. Brown to olive grey above (shining iridescent green/purple from some angles) with pale brown dorsolateral stripe, dark brown upper flanks, sometimes a weak pale midlateral stripe, dark head and barred lips. On breeding males, pattern reduced to absent, and limbs, tail, chin, throat flushed with red.

Variety of woodlands, heaths and vine thickets from Princess Charlotte Bay on CYP, through Torres Strait is. to sthn NG. Also ne. NT.

E. Vanderduys

Lygisaurus malleolus. Mt Lewis.

Red-tailed Litter-skink

Lygisaurus malleolus

SVL 32mm

Large disc in lower eyelid; round to slightly horizontally elongate ear opening surrounded by numerous sharp lobules; 7 upper labial scales. Brown with reddish orange flush on hindlimbs and tail, and reddish flush on throat of males. Tail virtually patternless ventrally.

Endemic. Open eucalypt woodlands in WT, EIU and sthn CYP, from Walkamin n. to Isabella Falls area, n. of Cooktown.

Lygisaurus parrhasius. Glennie Tableland.

Fire-tailed Litter-skink

Lygisaurus parrhasius

SVL 33mm

Dorsal scales with weak carinations, comprising series of small points; large disc in lower eyelid; round ear opening surrounded by long, pointed lobules. Black, with prominent white vertebral, dorsolateral and midlateral stripes, bright red tail. On older specimens pattern much weaker, with darker stripes, paler background, browner tail.

Endemic. Known only from sandstone habitats of Glennie Tableland, 60km nw. of Lockhart R. community on ne. CYP. Presumably rock-inhabiting.

E. Vanderduys

Lygisaurus rimula. Sir William Thompson Ra.

Crevice Litter-skink

Lygisaurus rimula

SVL 39mm

Dorsal scales with 4–5 weak carinations comprising series of small points; each scale with curved hind edge; large disc in lower eyelid; vertical ear opening surrounded by sharp lobules; dorsally depressed body. Copper brown, often with paler flecks, and prominent gold/silver dorsolateral stripe from above eye to base of tail. Outer edge of stripe sharp and straight, inner edge ragged, margining dark laterodorsal stripe. Upper flanks black, enclosing narrow, broken, silvery midlateral stripe, poorly defined on large males.

Endemic. Rock-inhabiting; mainly along creeks edged with vine thickets between Pascoe R. and Coen area, CYP.

Lygisaurus rococo. Chillagoe.

Chillagoe Litter-skink

Lygisaurus rococo

SVL 39mm

Large disc in lower eyelid; round ear opening with flat blunt lobules around margin; dorsally depressed body with long limbs. Coppery brown and patternless, iridescent metallic green sheen when viewed obliquely.

Endemic. Rock-inhabiting; limestone outcrops and granite boulders in Chillagoe and Undara districts, EIU.

C. Dollery

Lygisaurus sesbrauna. Captain Billy's Landing.

Eastern Cape Litter-skink

Lygisaurus sesbrauna

SVL 34mm

Small disc in lower eyelid; round ear opening with large, sharp lobules on anterior edge, smaller sharp lobules around margin; usually 7 upper labial scales; very small size. Reddish brown to yellowish brown above with dark vertebral stripe, 2–3 fine, dark paravertebral lines, finely dark-edged pale dorsolateral stripes, sometimes dark longitudinal lines on flanks. Breeding males lose pattern, develop red flush over tail and hindlimbs.

Endemic. Leaf litter in monsoon forests, open forests, woodlands, heaths on CYP, s. to Silver Plains.

Lygisaurus tanneri. Cooktown area.

Tanner's Litter-skink

Lygisaurus tanneri

SVL 37mm

Small disc in lower eyelid; round ear opening with ear-lobules absent or low and flat around margin; usually 7 upper labial scales; very small size. Pale to dark brown with obscure pale flecks above; usually darker on flanks, sometimes pale dorsolateral stripes, particularly on juv. Breeding males develop red flush on throat and tail.

Endemic. Riverine and monsoon forests between Endeavour R. and Starcke Wilderness, between se. CYP and nthn WT.

Sun-loving Litter-skink
Lygisaurus zuma
SVL 34mm

Large disc in lower eyelid; round to nearly round ear opening with low flat lobules around margin; usually 7 upper labial scales; very small size. Iridescent greyish brown with copper on head, weakly patterned with longitudinal rows of dark flecks on back, concentrating to form dark zone on upper flanks. Breeding males develop red throat and tail.

Lygisaurus zuma. Kirrama Ra.

Endemic. Open forests, riparian forests in CQC and WT between Mackay and Kirrama districts.

Genus *MAGMELLIA* Monotypic genus.

Orange-speckled Forest Skink
Magmellia luteilateralis
SVL 92mm

Well-developed limbs overlap when pressed to side of body, all with 5 digits; one lower labial scale contacts postmental scale; 3rd pair of chin shields separated by 5 longitudinal rows of smaller scales; scales along top of 4th toe in 2 rows with oblique sutures along almost entire length; movable lower eyelid lacking a transparent disc; no lobules on ear opening; parietal scales in contact; frontoparietals divided; smooth, shiny scales. Brown with a large black patch above forelimb, usually smaller dark patches on side of neck and above ear, burnt orange flanks marked with numerous, conspicuous small white spots.

Magmellia luteilateralis. Eungella NP.

R. Browne-Cooper

Endemic. Restricted to rainforest above 900m in Eungella NP, CQC. Lives in large rotting logs, often seen by day with head protruding from cavity.

Genus *MENETIA*

5 Aust. spp.; 2 occur in Qld. Four well-developed limbs with 4 fingers and 5 toes; lower eyelid fused to form large fixed spectacle; minute ear opening without lobules; parietal scales in contact; frontoparietals fused to form one shield; interparietal scale distinct; extremely small size.

Dry habitats, ranging from mulga woodlands of arid zones to dry sclerophyll forests within moister estn areas. Amongst Australia's smallest reptiles. With snout–vent lengths barely exceeding 20mm, some are no larger than medium-sized insects. *Menetia* are furtive leaf-litter inhabitants, active on the surface by day but seldom venturing far from cover and rapidly retreating under leaves if disturbed. Egglayers.

KEY TO *MENETIA*

a b

1 One large supraciliary scale contacts 1st supraocular scale **(a)** .... ***greyii***

2 Two supraciliary scales contact 1st supraocular scale **(b)** .... ***maini***

Common Dwarf Skink

Menetia greyii

SVL 38mm

Only 2 supraciliary scales, 1st very small and excluded from contacting 1st supraocular scale by much larger 2nd supraciliary; pale midlateral stripe. Brownish grey to grey with 2 or more series of dark dorsal dashes, and broad dark upper lateral zone above a white midlateral stripe.

Widespread over most bioregions; like all *Menetia*, shuns moist habitats, so occurrence in estn parts patchy, confined to dry or well-drained sites. Also dry parts of all mainland States.

Main's Dwarf Skink

Menetia maini

SVL 25mm

Only 2 supraciliary scales, 2nd not much larger than 1st and both contacting supraocular scale, excluding contact between 2nd supraciliary scale and prefrontal scale. Brown to greyish brown with dark flecks on each scale, sometimes vague pale dorsolateral line, dark brown to black flanks and sometimes indication of pale midlateral stripe.

Various dry shrubland, grassy woodland habitats in NWH, MGD, DEU, GUP. Also nthn NT and WA.

Menetia greyii. Currawinya NP.

Menetia maini. Mt Isa area.

Menetia greyii. Capella.

Menetia maini. Mt Isa.

Genus *MORETHIA*

8 Aust. spp.; 4 occur in Qld. Well-developed limbs each with 5 digits; lower eyelid fused to form fixed spectacle; parietal scales in contact; frontoparietals and interparietal fused to form one large shield; frontal scale much larger than each prefrontal. Throats of breeding males often flushed with red.

Favours dry conditions, occupying variety of wooded and shrubby habitats in the interior, and well-drained, often sandy soils or stony slopes along the humid e. coast and ranges. These swift, sun-loving skinks often forage on open ground or amongst rocks but seldom far from leaf litter or other low cover. In dry sthn and sw. areas, *Morethia* are the most common small diurnal skinks, occupying a niche largely exploited by *Carlia* further e. and n. Egglayers.

KEY TO *MORETHIA*

1 Pale dorsolateral stripe prominent and usually extending forward beyond eye; black upper lateral stripe extends forward to snout 2

Pale dorsolateral stripe weak to absent and not extending forward beyond eye; dark upper lateral stripe not extending forward beyond eye 3

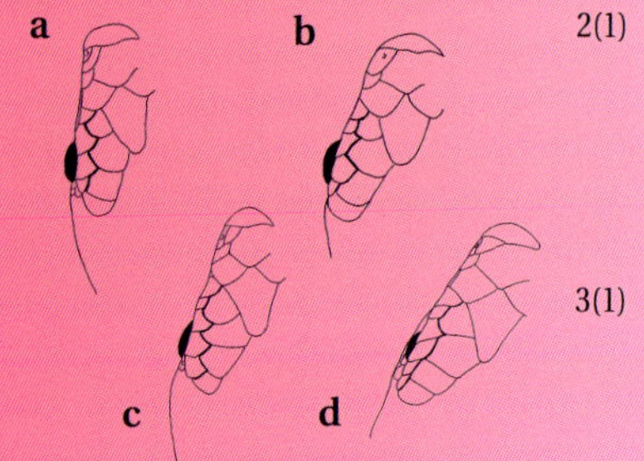

2(1) First 3 supraciliaries penetrate deeply between supraoculars to form ragged junction (**a**); only 3–5 scale rows between broad pale dorsolateral stripes .... ***ruficauda***

First 2 supraciliaries not penetrating deeply between supraoculars to form ragged junction (**b**); 6 scale rows between narrow pale dorsolateral stripes .... ***taeniopleura***

3(1) Subdigital lamellae sharply keeled; last 3 supraciliaries penetrate deeply between supraoculars (**c**) .... ***adelaidensis***

Subdigital lamellae smooth to bluntly keeled; last 3 supraciliaries form roughly straight line along junction with supraoculars (**d**) .... ***boulengeri***

Adelaide Snake-eyed Skink

Morethia adelaidensis

SVL 53mm

Usually 5 supraciliaries, last 3 largest and penetrating deeply between supraoculars; fragmented pattern.

Grey to olive brown with broken dark dorsal stripes, sometimes indistinct broken pale dorsolateral stripe, and dark brown upper flanks with irregular dark and pale flecks. Pale midlateral stripe (if present) wavy and irregular.

Chenopod shrublands, particularly those bordering saline flats, in CHC, ML. Also dry to arid sthn Australia.

Morethia adelaidensis. Currawinya NP.

Morethia boulengeri. Breeding male with red throat flush. Meandarra area.

Boulenger's Snake-eyed Skink

Morethia boulengeri

SVL 50mm

Usually 6 supraciliaries, 1st and 3rd largest, decreasing evenly in size from 3rd to 6th, and forming roughly straight-edged junction behind 1st supraocular. Coppery brown to greyish brown above with longitudinally aligned dark dashes, black upper flanks and broad, conspicuous white midlateral stripe. Tail of juv. flushed with red.

Lightly timbered areas, usually assoc. with heavy soils, in CHC, MGD, DEU, ML, BB, NET, SEQ, mainly w. of coastal ra. but nearing coast in Ipswich area. Widespread in dry parts of sthn Australia.

Fire-tailed Skink

Morethia ruficauda

SVL 46mm

Four supraciliaries, 1st 3 penetrating deeply between supraoculars to form ragged junction; bold colouration. Black with simple, sharp white dorsolateral stripes beginning on snout and midlateral stripes beginning on lips, all extending back to hips. Hips, hindlimbs and tail flushed with deep red.

Morethia ruficauda. Ethabuka.

Usually well-drained rocky areas; also desert dunes with spinifex. NWH, CHC. Also central and nw. Aust. Represented in Qld and over most of range by *M. r. ruficauda.* One other ssp. in parts of WA.

Morethia taeniopleura. Wivenhoe Dam.

Eastern Fire-tailed Skink

Morethia taeniopleura

SVL 42mm

Five supraciliaries, 3rd and 4th penetrating deeply between supraoculars. Rich brown to greyish brown with prominent, very narrow, pale dorsolateral stripe extending back from brow or nostril, black upper flanks extending back from snout and broad prominent white lower lateral stripe, all reaching back to hips. Hips, hindlimbs and base of tail flushed with reddish brown to rich red.

Endemic. Ranges and well-drained coastal areas in CYP, EIU, WT, DEU, BB, SEQ. Often associated with rock outcrops; pops on Moreton Bay is. occupy sandy heaths and woodlands.

Genus *NANGURA* Monotypic genus.

Nangur Spiny Skink

Nangura spinosa

SVL 95mm

Very strongly keeled dorsal scales, becoming spiny on tail; weakly keeled ventral scales; rough head scales; divided frontoparietal scales; widely separated parietals; movable lower eyelid with no transparent disc; large ear opening with no lobules. Brown with weak, irregular, dark dorsal bands and pale vertical bars on flanks.

Nangura spinosa. Nangur NP.

Endemic. Restricted to vine thickets on heavy soils in SEQ. Only 2 pops known, discovered in early 1990s at Nangur and Oakview NPs near Murgon and Kilkivan. Sedentary, rarely foraging, resting by day with head and forebody protruding from burrows under tree buttresses, surface roots or rocks on sloping terrain, to ambush passing invertebrates. Livebearers, probably producing small litters. Young share mother's burrow for up to 9 months; often perch beside her, or even on her back, at burrow entrance.

Genus *NOTOSCINCUS* 2 Aust. spp.; 1 occurs in Qld.

Ornate Snake-eyed Skink

Notoscincus ornatus

SVL 39mm

Well-developed limbs each with 5 digits; lower eyelid fused to form fixed spectacle, completely surrounded by granules; large ear opening with small anterior lobules; parietal scales in contact; frontoparietals fused to form one shield; interparietal free; subdigital lamellae divided or with 2 parallel keels. Coppery brown to bronze-brown above, with or without a vertebral row of black dots (nthn pop.), or 3 rows of black dots and/or dashes (sw. pops), upper lateral series of dark squarish bars and white midlateral stripe. Ill-defined ssp. *N. o. wotjulum* has broad, unbroken black upper lateral stripe.

Notoscincus ornatus ornatus. Mt Cooper Stn.

N. o. ornatus occupies sandy flats in GUP and EIU, and in sandy spinifex deserts of CHC. *N. o. wotjulum* recorded from NWH. Both ssp. also occur in NT and WA. Basks and forages among leaf litter and on adj. exposed ground, seldom venturing far from cover. The CHC pop. with more dorsal spots is probably linked to those from Central Aust. and may represent a separate sp. Egglayer.

Notoscincus ornatus wotjulum. Doomagee area.
S. Macdonald

Genus *OPHIOSCINCUS*

3 Aust. species; all occur in Qld. Limbs either completely absent or reduced to 4 minute stumps; very small eye with movable lower eyelid, lacking transparent disc; ear opening represented by scaly depression; blunt snout, tipped with waxy cuticle; long body with smooth glossy scales; orange-yellow ventral pigments.

Forests and heaths of SEQ. Also ne. NSW. These worm-like burrowers live in compost, in tunnels of invertebrates, and in loose soil under stones, logs and leaf litter. They require damp conditions and can desiccate rapidly if exposed to warm dry air. They feed on invertebrates, including earthworms. Egglayers; clutches of 2–3 eggs.

KEY TO *OPHIOSCINCUS*

a b c

1 Limbs present.. ***truncatus***
Limbs absent.. 2

2 Prefrontal scales present (**a**); 2 loreal scales (**b**); midbody scales in 20–24 rows.................... ***ophioscincus***
Prefrontal scales absent; one loreal scale (**c**); midbody scales in 18–19 rows..................... ***cooloolensis***

Cooloola Snake-skink
Ophioscincus cooloolensis
SVL 69mm

No limbs; no prefrontal scales; one loreal scale; midbody scales in 18–19 rows. Silvery grey above with 2–4 dorsal lines of black spots, black upper flanks and yellow to orange ventral surfaces, black beneath tail.

Endemic to SEQ. Coastal heaths, woodlands, rainforests on white sands

Ophioscincus cooloolensis. Cooloola NP.

at Cooloola and Fraser Is. Also disjunct record from Kroombit Tops, upland area nearly 300km to nw.

Ophioscincus ophioscincus. Mt Glorious.

Yolk-bellied Snake-skink
Ophioscincus ophioscincus
SVL 97mm

No limbs; prefrontal scales present; 2 loreal scales; midbody scales in 20–24 rows. Silvery grey above with 4–6 dorsal lines of black spots on body and 2–4 on tail, black upper flanks, yellow to orange ventral surfaces, black beneath tail.

Endemic. SEQ, n. to Bulburin SF near Many Peaks. Wet sclerophyll forests, rainforests, vine thickets, penetrating drier habitats via moist gullies and depressions.

Ophioscincus truncatus. Mary Cairncross Park.

Short-limbed Snake-skink
Ophioscincus truncatus
SVL 79mm

Limbs present but reduced to minute stumps; no contrasting black flanks. Brown to grey, darkening to black on tail, with narrow dark lines or rows of dashes on back and flanks, cream to orange ventral surfaces, black beneath tail.

SEQ, from Border Ra. n. to Conondale Ra. and Noosa. Also ne. NSW. Mainland pops normally occur in rainforest, wet sclerophyll forest on rich heavy soils at elevated sites, although reaching sandy coastal lowlands at Noosa. Moreton and North Stradbroke Is. pops occur in forests, woodlands, heaths on pale sandy soils.

Worm Skinks
Genus *PRAETEROPUS*

4 Aust. spp; all occur in Qld. Limbless and worm-like; body scales smooth and glossy; parietal scales in contact; lower eyelid scaly with no transparent disc; ear opening absent without trace or represented by a scaly depression; prefrontal scales absent (small and widely separated on one sp.); nasal scales large and in broad contact; snout round.

Mid-eastern to north-eastern Qld in dry forests and woodlands, vine thickets and rainforests. Burrowers, moving through friable soils, sand and compost under rocks, logs and leaf litter.

KEY TO *PRAETEROPUS*

1	Supraocular scales 2	2
	Supraocular scales 3	3
2(1)	Pattern comprises pale dorsum and contrasting darker flanks; prefrontal scales absent	***auxilliger***
	Pattern comprises dorsal and lateral rows of dark spots; prefrontal scales present	***gowi***
3(1)	Pattern comprises pale dorsum and contrasting dark flanks; 4 lower labial scales; upper elevations of Mt Abbott	***monachus***
	Pattern comprises dorsal and lateral rows of dark spots; 5 lower labial scales; mid-east Qld lowlands	***brevicollis***

Praeteropus auxilliger. Lake Elphinstone, Qld.

Praeteropus brevicollis. Capella region.

Sandstone Worm Skink
Praeteropus auxilliger
SVL 87mm

2 supraocular scales; 4 lower labial scales; 5 upper labial scales; no prefrontal scales; upper secondary temporal scale entire; frontoparietal scales separated; pale dorsal colour with contrasting dark flanks. Pale pink to yellowish brown above with weakly contrasting darker smudges on bases of scales. Flanks strongly contrasting dark brown to almost black.

Endemic. Lake Elphinstone and Redcliffe Tableland area in nthn BB. Recorded from eucalypt and acacia woodlands associated with sandstone.

M. Swan

Short-necked Worm Skink
Praeteropus brevicollis
SVL 83mm

Usually 3 supraocular scales; 5 lower labial scales; 6 upper labial scales; no prefrontal scales; frontoparietal scales almost always in contact; little contrast between dorsal and lateral surfaces. Brown above with darker head, tail, lower flanks, chin and throat, and obscure brown spots on bases of scales.

Endemic. BB, CQC, from about Cracow district n. to Finch Hatton, inland to Clermont. Variety of habitats: rainforest, vine thickets, dry sclerophyll forests.

Speckled Worm Skink
Praeteropus gowi
SVL 108mm

2 supraocular scales; 4 lower labial scales; usually 5 upper labial scales; prefrontal scales present

Praeteropus gowi. White Mountains NP.

but small and widely separated; upper secondary temporal scale divided into 2; frontoparietal scales in contact or separated; little contrast between dorsal and lateral surfaces. Brown to yellowish brown with darker head and tail-tip and prominent dark dashes, one per scale, arranged in longitudinal rows.

Endemic. Nthn DEU, EIU, from Magnetic Is. to White Mnts NP and n. to Mt Mulligan. Woodlands, evergreen vine thickets.

Praeteropus monachus. Preserved holotype. Mt Abbott.

Mount Abbott Worm Skink

Praeteropus monachus

SVL 72mm

3 supraocular scales; 4 lower labial scales; 6 upper labial scales; no prefrontal scales; upper secondary temporal scale entire; frontoparietal scales in narrow point contact; pale dorsal colour with contrasting dark flanks. In preservative, pale yellowish brown above with small darker blotches on scales. Flanks strongly contrasting dark brown to almost black, forming a series of narrow stripes. Dark spots on each ventral scale form broken lines.

Endemic. Known only from upper elevations of Mt Abbott in nthn BB, from open forest at 800m alt. and montane heath overlying rock pavements at 1,000m alt.

Genus *PROABLEPHARUS* 2 Aust. spp.; 1 occurs in Qld.

Slender Snake-eyed Skink

Proablepharus tenuis

SVL 32mm

Four small but well-developed limbs each 5 digits; large eye with lower eyelid fused to form a large fixed spectacle; minute ear-opening; parietal scales in contact; frontoparietal scales divided; interparietal scale free; plain pattern but when present comprises obscure dark pigment in centre of scales; very small size. Copper brown to olive with pattern absent or reduced to fine dark spot on most scales. Sides of head and neck on mature males flushed with red.

Tropical open forests, woodlands in EIU, BB, GUP, DEU, NWH. Also nthn NT, nthn WA. Normally occurs where groundcover dominated by spinifex, speargrass or leaf litter.

R. Browne-Cooper

Proablepharus tenuis. Townsville area.

Pygmy Skinks
Genus *PYGMAEASCINCUS*

3 Aust. spp.; all occur in Qld. Four short, widely spaced, well-developed limbs with 4 fingers and 5 toes; lower eyelid partially fused (an immoveable spectacle with a suture along upper edge); 10 or fewer scales along top of 4th toe; frontoparietal and interparietal scales fused to form a single shield; orange flush on tails of breeding males; extremely small size.

Estn Qld, in well drained habitats where dry leaf-litter accumulates at bases of trees and shrubs. Swift, secretive skinks that forage over and through leaf-litter and seldom venture into exposed areas. Egglayers.

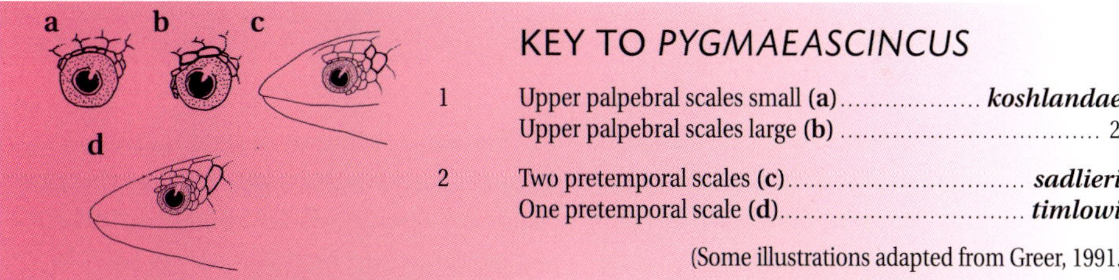

KEY TO *PYGMAEASCINCUS*

1 Upper palpebral scales small **(a)** .... ***koshlandae***
Upper palpebral scales large **(b)** 2

2 Two pretemporal scales **(c)** .... ***sadlieri***
One pretemporal scale **(d)** .... ***timlowi***

(Some illustrations adapted from Greer, 1991.)

E. Vanderduys

Pygmaeascincus koshlandae. Breeding male. Almaden area.

Pygmaeascincus sadlieri. Nelly Bay, Magnetic Is.

Koshland's Pygmy Skink
Pygmaeascincus koshlandae
SVL 28mm

Five supraciliary scales, with 3 contacting 1st supraocular scale; 3 supraocular scales; small upper palpebral scales; 2 pretemporal scales. Uniform brown above with darker flanks and pale lips with brown flecks. Breeding males develop orange flush over chin, throat and tail.

Endemic. WT and nthn EIU; from Palmer River and Shipton's Flat region to Mt Mulligan and Almaden areas.

Magnetic Island Pygmy Skink
Pygmaeascincus sadlieri
SVL 22mm

Five supraciliary scales, with 3 contacting 1st supraocular scale; 2 supraocular scales;

large upper palpebral scale; 2 pretemporal scales. Pale brown with pattern reduced to narrow, paler dorsolateral stripe and dark brown flanks.

Endemic. Known only from Magnetic Is. Recorded from woodlands, well-shaded gully and disturbed garden area.

Low's Pygmy Skink
Pygmaeascincus timlowi
SVL 29mm

Five supraciliary scales, with 3 contacting 1st supraocular scale; usually 3 supraocular scales, but these sometimes fuse to form 1–2 scales; large upper palpebral scales; one pretemporal scale. Dark brown to grey with obscure dark and pale streaks, reflecting shining green/purple iridescence from some angles. Breeding males develop reddish orange flush on tail.

Pygmaeascincus timlowi. Meandarra area.

Endemic. Wide variety of habitats, from closed semi-evergreen vine thickets to open woodlands and heath/spinifex associations on heavy to sandy soils in WT, EIU, DEU, BB, SEQ, e. to Helidon district.

Genus *SAIPHOS* Monotypic genus.

Three-toed Skink
Saiphos equalis
SVL 75mm

Very short, widely spaced limbs with 3 fingers and toes; movable lower eyelid with no transparent disc; ear opening reduced to scaly depression; frontoparietal scales divided; parietal scales in contact; very glossy texture. Brown with several narrow dark lines or rows of dashes on back, sharply contrasting darker flanks, bright yellow to bright orange ventral pigment from chest onto tail. Underside of last portion of tail black.

SEQ, BB. Patchily distributed in eucalypt forests, woodlands, vine thickets, occasionally rainforests n. to Kroombit

Saiphos equalis. Girraween NP.

Tops and w. to Carnarvon NP. Also estn NSW. Secretive, moisture-loving; dwells in soft soil under rocks, logs, leaf litter. Livebearer.

Shade Skinks
Genus *SAPROSCINCUS*

12 Aust. species; 11 occur in Qld. Four well-developed limbs, each with 5 digits (except *S. tetradactylus*, with only 4 fingers); movable lower eyelid enclosing transparent disc; frontoparietal scales divided; parietal scales in contact; interparietal distinct; 6 upper labial scales; no lobules on ear opening; weakly glossed to matt-textured body scales; pale spot on rear base of hindlimb.

Rainforests of SEQ, CQC, WT and sthn CYP. Each major forest block, isolated by broad tracts of drier terrain, supports its own suite of species. Also estn NSW, sthn Vic. Some larger *Saproscincus* often seen beside walking trails, basking amongst surface litter in weak dappled sunlight, or even roosting at night in vegetation up to 1m above ground. The smallest spp. are much more secretive and infrequently observed active on the surface. Egglayers; clutches of 2 eggs.

KEY TO *SAPROSCINCUS*

1	Four fingers	***tetradactylus***
	Five fingers	2
2(1)	Rear edge of parietal scales edged by 6 scales (**a**), a nuchal and 2 temporals on each side; WT to CYP	3
	Rear edge of parietal scales edged by 4 scales (**b**), a nuchal and one temporal on each side, SEQ to CQC; if a nuchal and 2 temporals on each side, then confined to CQC	6
3(2)	Snout pointed; maximum SVL only 34mm	***czechurai***
	Snout rounded; maximum SVL 42mm or more	4
4(3)	Back golden; flanks dark with golden spots; Cape Melville on CYP only	***saltus***
	Back shades of brown; sometimes dark upper flanks but without golden spots; WT only	5
5(4)	Paravertebral scales 46–50; maximum SVL 42mm; nthn WT, from Cape Tribulation northwards	***lewisi***
	Paravertebral scales 50–59; maximum SVL 49mm; sthn WT, from Roaring Meg southwards	***basiliscus***
6(2)	Mid-east	7
	South-east	8
7(6)	Tail short, on average only 130 per cent of SV; parietal scales usually edged by 6 scales	***hannahae***
	Tail long, about 145 per cent of SVL; parietal scales edged by 4 scales	***eungellensis***
8(6)	Supraciliary scales usually 6; back pattern weak to absent	9
	Supraciliary scales usually 7; back pattern often includes a mosaic of dark and pale scales	10
9(8)	Large pale blotches on dorsal surface of tail; 22 or more lamellae under 4th toe; no pale midlateral stripe; ventral pattern, if present, a scattering of pale brown spots	***challengeri***
	No pale blotches on dorsal surface of tail; 22 or fewer lamellae under 4th toe; a thin pale midlateral stripe; ventral pattern, if present, regular longitudinal rows of dark markings	***oriarus***
10(8)	Ventral surfaces of adult males bright lemon yellow; no sharp delineation in colour or pattern between dorsal and lateral surfaces of tail; no dark ventrolateral line on tail, at most a row of dark dashes	***spectabilis***
	Ventral surfaces of adult males very pale yellow; females have a bold rusty stripe delineating dorsal and lateral surfaces of tail; a narrow dark ventrolateral line on tail	***rosei***

Saproscincus basiliscus. Charmillan Creek.

Southern Wet Tropics Shadeskink

Saproscincus basiliscus

SVL 49mm

Midbody scales in 22–29 rows; paravertebral scales in 50–59 rows; 6 scales contacting rear edge of parietal scales (a nuchal and 2 temporals on each side). Shades of brown to reddish brown, with or without dark and pale flecks, sometimes dark-edged pale dorsolateral stripe on anterior body, usually broad dark upper lateral flush extending forward as stripe to nostril and back for varying distances along body.

Endemic. Rainforests of sthn WT, between Mt Elliot and Roaring Meg. Over parts of range it overlaps with either *S. tetradactylus* or *S. czechurai*, contrasting sharply with both these diminutive, secretive spp. by its larger size and more obvious surface activity. Excluding a very narrow overlap, it is replaced in nthn WT by *S. lewisi.*

Challenger's Shadeskink

Saproscincus challengeri

SVL 57mm

Very large size; usually 6 supraciliary scales; very long tail with several large, pale blotches on anterior dorsal surface; 4 scales contacting rear edge of parietal scales (a nuchal and a temporal on each side). Shades of brown, uniform or with scattered pale brown scales, and dark dorsolateral flecks tending to form irregular stripe, at least anteriorly. This extends back for various distances along body to hips and tail where it can develop rusty upper edge. Narrow dark ventrolateral stripe normally extends along tail, clearly defining dark lateral and pale ventral colours. Ventral surfaces cream (sometimes flushed with pale yellow on mature males; weak posterior flush on some adult females), sparsely to heavily marked with scattered brown spots. Up to 20 per cent of females plain brown with straight, prominent, pale-edged, dark dorsolateral stripe along body and tail.

SEQ, in rainforests of McPherson Ra. and hinterland, n. to Mt Tamborine and Cunningham's Gap. Also adj. NSW.

Saproscincus challengeri. Lamington NP.

Saproscincus czechurai. Charmillan Creek.

Saproscincus eungellensis. Eungella NP.

Czechura's Shadeskink

Saproscincus czechurai

SVL 34mm

Extremely small size; short tail; sharp snout; 6 scales contacting rear edge of parietal scales (a nuchal and 2 temporals on each side). Pale to dark reddish brown with small dark and pale flecks, sometimes coalescing to form wavy lines, and sometimes dark-edged, reddish brown dorsolateral stripe, usually most discernible on hips and tail. Breeding males become flushed with black and sharply spotted with white on sides of head and neck, and orange on hindlimbs and tail.

Endemic. Cool upland rainforests above 360m in WT, between Shipton's Flat and Cardwell Ra. Extremely secretive, living under stones, logs, leaf litter in moist sheltered sites.

Eungella Shadeskink

Saproscincus eungellensis

SVL 67mm

Long tail; large size; usually 6 supraciliary scales; 4 scales in contact with rear edge of parietal scales (a nuchal and temporal on each side); 26–34 subdigital lamellae under 4th toe. Males brown with scattered pale scales, sharply delineated from flanks by dorsolateral row of dark markings. Ventral surfaces with dark flecks scattered or arranged as broken rows. Females similar but darker, or plain brown with dark dorsolateral stripe, its upper edge pale, and paler grey to brown flanks.

Endemic. Rainforest above 700m, normally near streams, in Clarke Ra., CQC.

Saproscincus hannahae. Mt Blackwood.

Hannah's Shadeskink

Saproscincus hannahae

SVL 38mm

Small size; relatively short tail; 6 scales contacting rear edge of parietal scales (a nuchal and 2 temporals on each side). Brown to reddish brown above, paler on flanks, with broken longitudinal lines formed by dark spots on some scales. Upper and lateral surfaces clearly delineated by pale dorsolateral stripe from eye to forelimb, often breaking into row of dark spots on body. Lips barred.

Endemic. CQC, in rainforests from Mt Dryander near Proserpine s. to Sarina district. Favours shady, moist areas like creek edges; seldom ventures far from cover.

Saproscincus lewisi. Shipton's Flat.

Northern Wet Tropics Shadeskink

Saproscincus lewisi

SVL 42mm

Midbody scales in 22–24 rows; paravertebral scales in 46–50 rows; 6 scales contacting rear edge of parietal scales (a nuchal and 2 temporals on each side). Shades of brown with scattered dark flecks above, and uniform or with longitudinal rows of flecks on flanks. Upper and lateral surfaces often clearly delineated by dorsolateral stripe from eye to forelimb, breaking into row of dark spots on body and often with dark lower and pale upper edges.

Endemic. Rainforests of nthn WT, between Cape Tribulation and Mt Webb. Largely replaces *S. basiliscus* as conspicuous, active *Saproscincus* within its range, but overlaps narrowly along parts of sthn limits.

Coastal Shadeskink

Saproscincus oriarus

SVL 43mm

Very slender body and long slender tail; narrow white midlateral line; usually 6 supraciliary scales; 4 scales in contact with rear edge

H. Hines

Saproscincus oriarus. North Stradbroke Is.

of parietal scales (a nuchal and temporal on each side). Uniform brown above with narrow white midlateral stripe from snout to forebody, or breaking at about midbody and reaching base of hindlimb. Narrow dark ventrolateral stripe along tail, sharply delineating dark lateral and pale ventral surfaces. Belly either patternless, marked with several longitudinal rows of dark streaks, or with a dark ventrolateral stripe between forelimbs and hindlimbs.

Densely vegetated wetlands on North Stradbroke Island in SEQ. Also ne. NSW.

Rose's Shadeskink

Saproscincus rosei

SVL 64mm

Usually 7 supraciliary scales; large size; long tail marked with a thin dark ventrolateral line and (on females) a rusty dorsolateral stripe sharply defining dorsal and lateral surfaces; usually 4 scales in contact with rear edge of parietal scales (a nuchal and temporal on each side). Extremely variable, particularly females. Brown to reddish brown above, usually with mosaic of lighter and darker scales. Upper flanks generally darker, delineated from paler dorsal colours and merging gradually to paler lower flanks. Females tend to have prominent, sharp-edged reddish brown stripe from hips onto tail. Some also have uniform brown backs, sharply contrasting

Saproscincus rosei. Uncommon colour form, seen only on females. Border Ra.

Saproscincus rosei. Lamington NP.

plain dark flanks, prominent white midlateral stripe. Ventral surfaces cream, with pale yellow wash on posterior half of abdomen on mature males and sometimes a weak wash on females, and typically with regular, longitudinal rows of faint brown spots.

Rainforests, adj. wet sclerophyll forests in SEQ, from D'Aguilar, Conondale and Jimna Ra. s. to McPherson Ra. Also ne. NSW.

Saproscincus spectabilis. Neutral Bay, NSW.

Cape Melville Shadeskink

Saproscincus saltus

SVL 42mm

Extremely long limbs and digits; 6–7 supraciliary scales; 6 scales contacting rear edge of parietal scales (a nuchal and 2 temporals on each side). Golden brown with scattered dark flecks above, and flanks brown to dark brown regularly spotted with gold to yellow.

Endemic. Boulders associated with rainforest on Cape Melville, CYP. The unique site is isolated by boulder fields. Swift, agile, long-limbed skinks, able to move rapidly across rocks and leap between boulders when pursued.

C. Hoskin

Saproscincus saltus. Cape Melville.

Gully Skink

Saproscincus spectabilis

SVL 59mm

Usually 7 supraciliary scales; no abrupt delineation in colour or pattern between top and sides of tail; usually 4 scales in contact with rear edge of parietal scales (a nuchal and temporal on each side). Shades of brown above, normally with scattered pale and dark scales, and dark flecks aligned to form irregular dorsolateral stripe on anterior body, extending back for varying distances to tail. Flanks similar to back or significantly darker, often with

Saproscincus tetradactylus. Curtain Fig NP.

pale midlateral stripe. Dorsal and lateral surfaces of tail ill-defined, with no sharp change in colour or pattern. Ventral surfaces white to bright lemon yellow, with dark flecks which align on boldly marked lizards to form series of roughly longitudinal streaks.

Rainforests, moist gullies on Mt Tamborine and Mc Pherson Ra., SEQ. Also estn NSW.

Four-fingered Shadeskink
Saproscincus tetradactylus
SVL 33mm

Only 4 fingers; extremely small size; short tail; sharp snout; 6 scales contacting rear edge of parietal scales (a nuchal and 2 temporals on each side). Shades of brown to dark reddish brown with iridescent sheen from some angles. Otherwise pattern is drab: several series of black longitudinal dashes, narrow reddish brown dorso-lateral stripe with broken dark edges, often an obscure dark W-shaped mark on rear of head between eyes, barred lips. Ventral surfaces pale lemon yellow.

Endemic. Rainforests of WT, from Macalister Ra. s. to Mt Spec. Very secretive, rarely foraging far from leaf litter and other groundcover in shaded moist areas.

Genus *SUPPRESSASCINCUS* Monotypic genus.

Cape York Worm Skink
Suppressascincus pluto
SVL 76mm

Limbless and worm-like; body scales smooth and glossy; lower eyelid scaly with no transparent disc; ear opening absent, represented by a scaly depression; nasal and 1st upper labial scale completely fused; 2 supraocular scales; parietal scales in contact; frontoparietal scales separated; opaque cuticle on snout. Brown, darker on tail,

Suppressascincus pluto. Heathlands.

with only paler edges to scales. Ventral surfaces pale brown.

Endemic. Ne. CYP, on sandy soils supporting monsoon forests, heaths, open forests and vine thickets, often associated with active termite mounds and termite-infested wood.

Genus *TECHMARSCINCUS* Monotypic genus.

Bartle Frere Skink
Techmarscincus jigurru
SVL 70mm

Long limbs, each with 5 long slender digits; movable lower eyelid enclosing large transparent disc; no lobules on ear opening; parietal scales in contact; frontoparietals divided. Coppery brown above with prominent black and cream dots, irregular broken pale dorsolateral stripe, black flanks and sometimes broken white midlateral stripe.

Endemic. Restricted to mist-enshrouded granite outcrops, stunted forests above 1440m on Mt Bartle Frere in WT, an island of year-round cool temperatures surrounded by warmer tropical rainforests. Basks on exposed rocks, sheltering in crevices and under slabs. *Techmarscincus* appears to be a northern outlier related to the more temperate genera *Acritoscincus* (restricted in Qld to NET), *Niveoscincus* and *Pseudemoia* (both confined to sthn States). Egglayer.

Techmarscincus jigurru. Mt Bartle Frere summit.

Blue-tongued Skinks; Shingleback
Genus *TILIQUA*

7 spp. in Aust., NG and estn Indonesia; 3 occur in Qld. Short limbs each with 5 short digits approx. equal in length (4th toe not markedly longer); movable lower eyelid lacking transparent disc; usually no lobules on ear opening; parietal scales separated; frontoparietals paired; row of scales between eye and upper labial scales; extremely large size.

Throughout Qld and all other States. Amongst the most familiar reptiles, due to their size (arguably world's largest skinks), distinctive appearance and ability to thrive in highly modified urban environments. Slow-moving, relying on bluff rather than speed to elude predators. When cornered, the mouth is gaped and the broad blue tongue protruded, accompanied by a loud hiss. Omnivorous, consuming flowers, berries, fungi and soft foliage, eggs, invertebrates and any small vertebrates slow enough to capture.

Livebearers.

Tiliqua multifasciata. Nappa Merrie Stn.

Centralian Blue-tongue
Tiliqua multifasciata
SVL 289mm

Smooth scales; short tail; one supraocular scale contacts frontal scale; numerous narrow bands. Pale grey to greyish brown with about 9–14 orange to yellowish bands across body, 8–10 on tail, prominent broad black stripe behind eye.

Spinifex grasslands, dry open shrublands in arid NWH, MGD, CHC, wstn ML. Also arid zones of all mainland States except Vic.

Shingleback; Stumpy-tail
Tiliqua rugosa
SVL 260–310mm

Extremely large, rough scales; broad triangular head; short bulbous tail. Very dark grey to greyish brown or almost black with sparse, irregular pale blotches or spots.

Dry woodlands, shrublands of CHC, ML, sthn MGD, BB, extending e. to Toowoomba. Also dry areas across sthn Aust. This distinctive skink can be confused with no other. Pairs meet each spring, normally rejoining the same partner. Estn Aust. pops are referred to as *T. r. aspera.* This and other ssp. occur in WA.

Tiliqua rugosa. Roma area.

Tiliqua scincoides scincoides. Carseldine.

Eastern Blue-tongue; Northern Blue-tongue
Tiliqua scincoides
SVL 300–320mm

Smooth scales; relatively long thick tail; 2 supraocular scales contact frontal scale; broad bands. Extremely variable, ranging through shades of grey to brown, with about 6–9 pale bands between nape and hips and 7–10 on tail. Ssp. *T. s. intermedia* larger, with bands tending to fragment into variegations or mottling dorsally but remaining distinct on flanks.

Throughout Qld. Scarce and patchy in arid zones of CHC, wstn MDG. All mainland states. Occupies broad range of habitats, from gardens and vacant allotments of inner Brisbane to mulga woodlands of sthn interior, coastal dunes, rainforest edges, tropical woodlands. *T. s. intermedia* is nthn, from CYP to nthn NT, WA. *T. s. scincoides* occupies remainder.

R. Valentic

Tiliqua scincoides intermedia. Camooweal area.

DRAGONS

The Condamine Earless Dragon (*Tympanocryptis condaminensis*) was described in 2014. It is restricted to a small area of highly modified habitat comprising cotton, sorghum, maize and millet crops and relict grassland along roadside verges in the Darling Downs, southern Brigalow Belt. It is listed as an endangered species.

Family **Agamidae**

Dragons have alert, upright postures, well-developed limbs, long tapering tails and rough skins. Many are adorned with crests of spines, dewlaps and erectable beards or frills. Most are extremely swift, dashing on all fours or raising their bodies to sprint on their hindlimbs.

All are diurnal and the primary sense they employ is vision. From an elevated perch on a stump or rock, or from a clear sandy patch between spinifex clumps, dragons keenly scan the skies for raptors and the surrounding terrain for the movements of potential prey, mates or rivals. Most feed on small invertebrates, captured with a dab of the short sticky tongue. Many also take occasional herbage, with the proportion of vegetation consumed increasing in larger dragons. Egglayers, depositing their clutches in burrows, usually excavated in open areas.

Dragons occur statewide, their diversity greatest in deserts and tropical woodlands. They fare poorly in rainforests, though 2 of the most spectacular species, the rainforest dragons (*Lophosaurus*), occupy widely separated northern and southern forest blocks. One of our largest and best-known lizards, the Water Dragon (*Intellagama lesueurii*), thrives alongside urban waterways in the heart of Brisbane and other eastern towns and cities.

KEY TO GENERA

1 A bulbous hump with 2 large spines on neck; large thorn-like spines over eyes, body, limbs and tail .... ***Moloch***
No bulbous hump on neck or large spines over eyes; spines present or absent on body but never large and thorn-like 2

2(1) A large, loose frill around neck .... ***Chlamydosaurus***
No large frill around neck 3

a

3(2) A row of enlarged scales curving under eye to above ear **(a)** .... ***Ctenophorus***
No row of enlarged scales under eye 4

4(3) Numerous spiny scales scattered over back 5
Spiny scales absent or, if present, arranged in one or more clear longitudinal rows including along vertebral line 6

5(4) Ear exposed; long slender spines along flanks .... ***Pogona***
Ear hidden; no long slender spines along flanks .... ***Tympanocryptis***

6(4) Scales along vertebral line enlarged 7
Scales along vertebral line not significantly larger than adjacent scale rows .... ***Diporiphora*** (part)

b

7(6) Tail strongly laterally compressed .... ***Intellagama***
Tail round in cross-section 8

8(7) Body moderate; femoral and preanal pores usually present **(b)** 9
Body strongly laterally compressed including high vertebral ridge; no femoral and preanal pores 13

9(8) Dorsal body scales (excluding enlarged vertebral row) small and uniform 10
Dorsal body scales include additional enlarged longitudinal rows 12

10(9) Keels on dorsal body scales parallel with midline .... ***Lophognathus***
Keels on dorsal body scales converging back towards midline 11

11(10) Pale dorsal stripe discontinuous or narrowly continuous with pale stripe along lower jaw .... ***Gowidon***
Pale dorsal stripe broadly continuous with pale stripe along lower jaw .... ***Tropicagama***

c

d

12(9) Scales on upper thighs grade back evenly from large to small **(c)**; keels on dorsal body scales parallel with midline .... ***Diporiphora*** (part)
Scales on upper thighs a contrasting mix of large and small **(d)**; if grading back evenly from large to small, then keels on only inner-most dorsal scales parallel to midline but most of remainder converge back towards midline .. ***Amphibolurus***

13(8) Gular fold present; brow angular; tail sharp-tipped; moist estn forest .... ***Lophosaurus***
No gular fold; no angular brow; tail blunt-tipped; nw woodlands .... ***Chelosania***

Genus *AMPHIBOLURUS*

4 spp. in Aust; 3 occur in Qld. Low spiny nuchal and vertebral crest; 1–2 dorsal crests on each side, and additional spines on sides of neck and sometimes flanks; exposed ear; long limbs and tail. Grey to black with 2 broad pale dorsal stripes or series of blotches,

often sharply and strikingly defined on mature males.

Semi-arboreal, perching on trunks and lower limbs in dry forests and along tree-lined water-courses. All are alert, and run on their hind limbs. They frequently display with arm-waving, head-bobbing and tail-lashing, particularly after short bursts of movement.

KEY TO *AMPHIBOLURUS*

1	Scales on upper thighs grade back evenly from large to small (**a**); NWH and CHC	***centralis***
	Scales on upper thighs a contrasting mix of large and small (**b**); widespread over interior and SEQ	2
2	Slender spines on flanks; mouth-lining yellow	***muricatus***
	No spines on flanks; mouth-lining pink	***burnsi***

Burns' Dragon
Amphibolurus burnsi
SVL 135mm

Usually 5 crests – well-developed nuchal and vertebral crest, with 2 additional dorsal crests on each side; no spines on flanks; mix of large and small scales on thigh with large spiny row along rear of thigh; pink mouth-lining. Grey to almost black with 2 broad pale stripes (sometimes deeply notched or broken to form blotches) from ear to hips, discontinuous with pale stripes from side of neck along lower jaw to chin.

Amphibolurus burnsi. Bendidee SF.

Dry woodlands and eucalypt-lined water-courses in BB, ML, CHC, DEU and MGD. Also adj. NSW, SA.

Centralian Tree Dragon
Amphibolurus centralis
SVL 122mm

Usually 5 crests – well-developed vertebral crest and 1–2 dorsal crests on each side; 2–3 longitudinal series of enlarged spinose scales on each side at rear of head; scales on upper thighs grade back evenly from large to small; keels on inner-most dorsal scales parallel to midline, those on outer crests parallel to or converging back towards midline, and remaining small dorsal scales have keels converging on midline. Brownish with broad pale dorsolateral stripe with undulating or

Amphibolurus centralis. Noonbah Stn.

notched upper edge, extending forward to side of neck, discontinuous with broad pale stripe from lower jaw back to postauricular fold.

NWH and CHC. Also central to sthn NT. Recorded from woodlands and on trees along water courses.

Amphibolurus muricatus. Girraween NP.

Jacky Lizard
Amphibolurus muricatus
SVL 120mm

Five crests – well-developed nuchal and vertebral crest, with 2 additional dorsal crests on each side; spines present on flanks; mix of large and small scales on thighs; bright yellow mouth-lining. Grey with 2 paler grey to white wavy to straight-edged stripes, often broken into series of blotches.

Dry sclerophyll forests, heaths, woodlands in NET, SEQ, sthn BB. Also estn NSW, Vic, se. SA.

Genus *CHELOSANIA* Monotypic genus.

Chameleon Dragon
Chelosania brunnea
SVL 118mm

Laterally compressed body; low nuchal and vertebral crest, all other scales small and uniform; short limbs; blunt-tipped tail; prominent dewlap; granular conical eyelids with small apertures rather like a chameleon. Grey to brown with darker streaks and variegations, broad dark bands on tail, narrow lines radiating from eye.

Recorded from woodlands dominated by *Eucalyptus platyphylla* in far nthn NWH. Also tropical woodlands across nthn NT, WA. Poorly known and rarely seen. Slow-moving, extremely secretive. Arboreal, but most specimens encountered crossing roads.

S. Macdonald

Chelosania brunnea. Doomadgee area.

Genus *CHLAMYDOSAURUS* Monotypic genus.

Frilled Lizard; Frillneck

Chlamydosaurus kingii

SVL 258mm

Distinctive frill of scaly skin which at rest lies folded over neck and shoulders; when erected, with mouth agape, encircles head and neck to dramatically alter appearance. Shades of grey with darker marbling over back and limbs. Frills of Qld lizards patterned mainly with grey to yellow; those across nthn Aust. often flushed with red or orange.

Tropical to subtropical woodlands in NWH, GUP, CYP, EIU, WT, BB, CQC, SEQ, s. to Greenbank. Also sthn NG, nthn NT, WA. Common in Brisbane area until 1960s; pops from s. of about Gympie have since declined dramatically and are probably endangered. Arboreal, mainly descending during wet season to rest on lower trunks of rough-barked trees or forage on ground. Frill, erected when threatened and during social encounters, serves to abruptly increase apparent size and may also have thermoregulatory function.

E. Vanderduys

Chlamydosaurus kingii. Defensive pose. Prince of Wales Is.

Chlamydosaurus kingii. Kippa Ring.

Genus *CTENOPHORUS*

35 Aust. spp.; 6 occur in Qld. A row of enlarged scales curving under each eye; sometimes a weak, spiny nuchal crest and spines on sides of neck, but no other spines; exposed ear; femoral and preanal pores present; some black ventral pigment on mature males.

Dry to arid inland areas. Spp. vary significantly in behaviour and habitat selection. Military Dragons are extremely swift, wholly terrestrial inhabitants of sandy deserts, foraging in open spaces and sheltering under spinifex; Ring-tailed Dragons perch on boulders and hide in crevices; Netted and Painted Dragons perch on stumps, stones and termite mounds and excavate shallow sloping burrows under bases of shrubs in various arid shrublands.

KEY TO *CTENOPHORUS*

1 Hindlimbs long, reaching eye or beyond when pressed to side of body 2
Hindlimbs short, reaching no further than tympanum 4

2(1) All scales small and uniform; no indication of spines on sides of neck or enlarged scales along vertebral line 3
Prominent to weak enlarged spines on sides of neck; enlarged scales along vertebral line, at least on nape .... ***slateri***

3(2) Pores extend along full length of thigh and curve sharply forward towards midline (**a**); extensive black ventral pigment on males .... ***isolepis***
Pores extend along up to three quarters of thigh, straight or curving slightly towards midline (**b**); black ventral pigment on males largely confined to throat and chest .... ***fordi***

4(1) Prominent black patch on side of neck .... ***clayi***
No prominent black patch on side of neck 5

5(4) Dorsal scales uniform in size; row of enlarged scales below eye prominent and projecting .... ***pictus***
Dorsal scales include roughly transverse, widely spaced rows of larger scales; row of enlarged scales below eye moderate and not projecting .... ***nuchalis***

S. Macdonald

Ctenophorus clayi. Ethabuka.

Black-collared Dragon
Ctenophorus clayi
SVL 58mm

No spines or crests; no enlarged vertebral scales; well-developed row of scales from under eye to above ear; short limbs; round head with short snout; black patch on side of neck. Fawn to reddish brown with dark reticulum, transverse rows of pale spots, narrow pale vertebral stripe and prominent black patch on each side of neck, sometimes extending around throat to form collar.

Sand dunes with spinifex at Ethabuka, CHC. Also sandy areas with spinifex in NT and WA.

Mallee Military Dragon
Ctenophorus fordi
SVL 58mm

No crests or spines; weakly discernible row of scales from under eye to above ear; femoral and preanal pores arranged in straight line along no more than three-quarters of thighs and curving weakly forward at midline; long limbs. Brown to reddish brown with prominent pale dorsolateral and lateral stripes, dorsolaterals edged above and below with black blotches. Ventral surfaces of males marked with black dappling on throat, black bar

Ctenophorus fordi. Wombah Gate area.

R. Valentic

on anterior chest and leading edge of forelimbs.

Sandy deserts of CHC. Also arid zones of sthn SA and WA, though status uncertain and likely a complex of spp. Extremely swift and alert, dashing on all fours across open sandy areas between clumps of spinifex or canegrass. Studies of allied sp. in NSW suggest annual pop. turnover.

Ctenophorus isolepis. Male. Windorah.

Central Military Dragon

Ctenophorus isolepis

SVL 70mm

No crests or spines; weakly discernible row of scales from under eye to above ear; femoral and preanal pores along full length of thighs, curving sharply forwards at midline; long limbs. Reddish brown, with dark-edged cream spots and pale dorsolateral stripes extending back from behind eye to hips, often breaking into blotches at midbody. Mature males have black flanks, extensive black ventral pigment.

Arid zones with spinifex on relatively bare ground in CHC, wstn MGD and NWH. One of most conspicuous desert lizards, particularly on sandy flats between dunes. Completely terrestrial, rarely perching on elevated sites. Dashes in short sprints between spinifex clumps. Represented in Qld by ssp. *C. i. gularis.* Other ssp. in NT, SA, WA.

Ctenophorus nuchalis. Quilpie area.

Central Netted Dragon

Ctenophorus nuchalis

SVL 115mm

Small spiny nuchal crest but no enlarged vertebral scales; well-developed row of scales from under eye to above ear; small body-scales with widely spaced rows of larger scales; some spines on sides of neck; round 'golfball-like' head; pores arranged in curve sweeping forward on anterior thigh; short limbs. Brown to reddish brown with fine dark network over back and sides, excluding only narrow pale vertebral stripe.

Semi-arid to arid shrublands in ML, CHC, MGD, DEU, NWH. Also arid parts of NSW, SA, NT, WA. Perches conspicuously on low rocks, stumps, termite mounds. When pursued, dashes unerringly in straight line to short sloping burrow at base of shrub or stump.

Painted Dragon

Ctenophorus pictus

SVL 65mm

Small spiny nuchal crest; vertebral row of enlarged keeled scales; some spines on sides of neck; very well-developed row of scales from under eye to above ear; short limbs. Brown to orange

Ctenophorus pictus. Female. Currawinya NP.

Ctenophorus pictus. Male. Nappa Merrie Stn.

with dark vertebral stripe and dark-edged pale bars, blotches or spots. On breeding males, lower lips, throat and limbs flushed with blue, anterior chest and shoulders bright yellow to orange.

Semi-arid to arid chenopod shrublands, sandy slopes of dunes and margins of saltpans in ML, CHC, MGD, DEU. Also arid zones of all mainland States. Perches on low stumps and bushes; excavates short burrow at base of shrubs.

Slater's Ring-tailed Dragon

Ctenophorus slateri

SVL 100mm

Prominent nuchal and vertebral crest; weakly discernible row of scales from under eye to above ear; some spines on sides of neck; moderately long limbs. Reddish brown with paravertebral rows of large dark spots, sometimes alternating with thin pale bands or transverse rows of small pale spots. Tail banded. Body pattern tends to weaken on adults. Mature males have more conspicuous tail bands and large dark patch on chest.

Rocky ranges, outcrops in NWH, MGD, CHC. Also NT, WA. Alert, basking on protruding rocks, dashing to shelter of rock crevice or burrow beneath larger boulder if approached.

Ctenophorus slateri. Lawn Hill NP.

Genus *DIPORIPHORA*

28 Aust. spp.; 11 occur in Qld. Tympanum exposed; dorsal crests and other spines usually absent, but 5 dorsal rows of enlarged spinose scales on 2 spp. and restricted on some others to just a few small spines behind ear; femoral pores usually absent, but 3–6 present on each side on 2 spp.; preanal pores usually present, up to 6 on each side. Pattern usually includes pale vertebral and/or dorsolateral stripes overlaying dark transverse bars, and often a dark shoulder blotch.

Dry forests, woodlands and heaths of e. and n., and spinifex grasslands of interior. They are largely terrestrial but frequently bask on shrubs, low fallen timber and spinifex clumps. Compared to other small dragons, *Diporiphora* tend not to be particularly swift, scuttling on all fours to cover of thick vegetation when pursued.

Identification can be problematic.

KEY TO *DIPORIPHORA*

1	Body scales all homogeneous with dorsals grading evenly to lateral scales	2
	Dorsal scales heterogeneous with rows of spines; or with some longitudinal rows along outer edge of paravertebrals either significantly larger than vertebral row or with raised hind edges; or dorsal scales not grading evenly to homogeneous laterals	6
2(1)	Ventral surfaces without stripes or longitudinally aligned mottling	3
	Ventral surfaces striped or with longitudinally aligned mottling	5
3(2)	Gular fold present	***lalliae***
	Gular fold absent	4
4(3)	Granular scales over arm but not to sides of neck; far nthn border region with NT	***magna***
	Granular scales over arm and along full length of scapular fold; estn Gulf of Carpentaria to interior of Cape York	***carpentariensis***
5(2)	Postauricular fold present with spines; 4 dark ventral stripes	***ameliae***
	Postauricular fold weak, without spines; 2 dark ventral stripes	***winneckei***
6(1)	3–5 dorsal crests of enlarged spinose scales	7
	No dorsal spines, but some longitudinal rows along outer edge of paravertebrals significantly larger than vertebral row or with raised hind edges	8
7(6)	Arc of spines in front of tympanum; no scattered spines on upper flanks	***nobbi***
	No arc of spines in front of tympanum; scattered spines present on upper flanks	***phaeospinosa***
8(6)	Scales in axilla granular	***granulifera***
	Scales in axilla not granular	9
9(8)	Gular fold present (**a**)	10
	Gular fold absent	***jugularis***
10(9)	Two upper canine teeth on each side; far nthn border region with NT	***sobria***
	One upper canine tooth on each side; widespread estn Qld to estn Gulf	***australis***

A. Emmott

Diporiphora ameliae. Breeding male. Valetta Stn.

Diporiphora ameliae. Valetta Stn.

Amelia's Spinifex Dragon

Diporiphora ameliae

SVL 67mm

No spinose dorsal crests; dorsal scales homogeneous – roughly equal in size; gular fold absent; postauricular fold present with low spines; scapular fold present. Male is brown with grey vertebral zone, yellow dorsolateral stripes and 8 dark bars. Prominent red flush on base of tail when breeding. Ventral surface with 4 grey stripes or longitudinally aligned mottling on body, and 3 V-shaped markings converging anteriorly on chin and throat. Female has grey dorsolateral stripes, more prominent dorsal bars and prominent dark ventral pattern.

Endemic. Spinifex on hard gravelly soils on dissected ridges of Goneaway Tablelands in MGD. Recorded from Noonbah and Valetta Stns and Bladensberg NP.

Tommy Roundhead Dragon

Diporiphora australis

SVL 50mm

No spinose dorsal crests; dorsal scales mostly homogeneous but prominent keels form ridges and vertebral and dorsolateral rows raised and dorsolateral row sometimes enlarged; lateral scales heterogeneous with scattered enlarged scales; gular fold present, with additional folds along shoulder and behind ear. Grey to yellowish brown or reddish brown, patternless or with about 6–7 grey to dark brown bars on body, overlaid by pale grey vertebral and cream to yellow dorsolateral stripes.

Heaths, dry forests, woodlands of SEQ, BB, CQC, estn DEU, EIU, WT and estn GUP. Also ne. NSW.

Diporiphora australis. Bowling Green Bay NP.

E. Vanderduys

Diporiphora carpentariensis. Littleton NP.

Diporiphora granulifera. Lawn Hill NP.

Gulf Two-lined Dragon

Diporiphora carpentariensis

SVL 68mm

No spinose dorsal crests; dorsal scales homogeneous, strongly keeled; no gular fold, but scapular fold strong; postauricular fold weak to strong with cluster of small spinose scales incl. one larger cream coloured spine; granular scales in axilla extend over arm and along full length of scapular fold. Pattern strong to weak. When present, incl. pale dorsolateral stripes overlying 8–9 dark bands, off-set either side of grey vertebral stripe, and dark blotch on axilla. Weak patterns retain dorsolateral stripes.

Endemic. Savannah woodlands and grasslands in estn GUP and central and wstn CYP.

Granulated Two-lined Dragon

Diporiphora granulifera

SVL 68mm

No spinose dorsal crests; dorsal scales homogeneous, though with dorsolateral row raised; no gular fold; postauricular fold strong to weak with little or no indication of spinose scales; scapular fold strong; granular scales in armpit extending onto neck. Brown with strong to weak pattern. Patterned individuals have pale dorsolateral stripes overlying 6–8 dark bands slightly offset on either side of broad greyish vertebral stripe. Plain individuals retain pale dorsolateral stripes. Granular scales in armpit are dark brown. Breeding males have pink flush on base of tail and large black patch in armpit.

Tropical savannah woodlands with spinifex, typically on rocky terrain in NWH. Also adj. NT.

Black-throated Two-pored Dragon

Diporiphora jugularis

SVL 68mm

No spinose dorsal crests; dorsal scales homogeneous but with vertebral, paravertebral and dorsolateral ridges; lateral scales heterogeneous with scattered larger scales; no gular postauricular and scapular folds; scales in armpits small but not granular. Pattern variable. Sub-adults and female

Diporiphora jugularis. Cooktown area.

E. Vanderduys

Diporiphora jugularis. Mature male. Iron Range.

often brown to reddish brown with pale dorsolateral stripes and about 4–7 dark brown bands broken by pale vertebral stripe. Adult male usually weakly patterned with cream dorsolateral stripes flecked flanks and dark brown or black band over chin and throat or dark spot on either side of neck.

Endemic. Estn CYP and Torres Strait, in tropical woodlands, shrublands and heaths.

Northern Deserts Dragon
Diporiphora lalliae
SVL 62mm

No spinose dorsal crests; dorsal scales homogeneous and strongly keeled; gular fold present, and scapular and postauricular folds strong; scales in armpit small but not granular. Strongly patterned individuals have 5–6 broad, dark bands, intersected by a wide grey vertebral stripe and narrow pale dorsolateral stripes, a prominent white stripe often extending from behind eye to above ear, and sometimes a dark spot over shoulder. Plain individuals may have just a few dark specks, and sometimes pale

Diporiphora lalliae. Tennant Creek area, NT.

dorsolateral stripes. Breeding males often develop a pink flush on sides of tail.

Far wstn NWH on NT border. Also across nthn arid zones of NT and WA, in woodlands and grasslands, typically incl. spinifex.

Diporiphora magna. Lake Argyle area, WA.

Yellow-sided Two-lined Dragon
Diporiphora magna
SVL.77mm

No spinose dorsal crests; dorsal scales homogeneous – roughly equal in size; no gular fold but postauricular and scapular folds strong. Yellowish brown to olive green. Some adults largely patternless except for large dark patch on anterior flanks. Juv. and subadult usually clearly marked with narrow, angular dark brown bars on body, overlaid by pale grey vertebral and narrow white dorsolateral stripes.

Tropical woodlands in far wstn NWH. Also nthn NT and WA.

Diporiphora nobbi. Girraween NP.

B. Schembri

Diporiphora sobria. Litchfield NP, NT.

Common Nobbi Dragon

Diporiphora nobbi

SVL 84mm

5 regular dorsal crests – well developed nuchal and vertebral crest, with 2 additional dorsal crests on each side; no enlarged spines on dorsolateral area; an arc of spiny scales enclose a small area behind ear. Pale or dark grey to brown with 2 very prominent white to yellow dorsal stripes. On breeding males, sides of tail-base often flushed with mauve to red and dorsal stripes bright yellow.

Dry forests, heaths and woodlands in NET, SEQ, BB, CQC, DEU, EIU and WT. Also NSW, nw. Vic and se. SA, Generally larger, more robust and swifter than most *Diporiphora*.

Northern Savannah Two-pored Dragon

Diporiphora sobria

SVL 69mm

Dorsal scales heterogeneous; 2 dorsolateral rows of raised pale scales, with large strongly keeled scales between them and small keeled scales on outer rows; robust build; gular fold present; postauricular fold weak to strong; scapular fold present but variable. Brown to grey, with pattern absent or diffuse, comprising dark dorsal bars and a dark patch on shoulder. Breeding males have strongly contrasting charcoal black, white and chestnut or orange-red on head, body and sometimes tail.

Far wstn NWH on NT border. Also across nthn NT and WA, in savannah woodlands and grasslands, often associated with rocky areas.

Diporiphora phaeospinosa. Blackdown Tablelands.

Black-throated Nobbi Dragon

Diporiphora phaeospinosa

SVL 72mm

5 irregular dorsal crests – well developed nuchal and vertebral crest, with 2 additional dorsal crests on each side; enlarged spinose scales scattered randomly over dorsolateral area; no arc of spiny scales behind ear, but a row of enlarged spines extends back from tympanum. Greyish brown with a pair of prominent pale dorsolateral stripes, and a series of

angular dark blotches on either side of midline. Breeding males develop a deep red flush on tail, brightest on base, and a black throat and chest.

Endemic. Moist eucalypt forests, woodlands and heaths among sandstone outcrops and plateaux in BB, in the vicinity of Carnarvon and Bigge Ra. areas and Blackdown Tablelands.

Canegrass Dragon; Blue-lined Dragon
Diporiphora winneckei
SVL 64mm

Dorsal scales homogeneous, roughly equal in size with no enlarged dorsolateral row; gular and postauricular folds weak without spines; scapular fold strong; very slender body and limbs. Pale bluish grey to yellowish brown with broad grey vertebral stripe, narrow white to yellow dorsolateral stripes and dark dorsal bars reduced to 6–8 paired blotches. Ventral surfaces smooth and silky white with 2 prominent stripes – silver-grey on males, and grey and yellow on females – converging at chest and pelvis.

Sand ridge deserts of CHC. Also adj. ne. SA and se. NT. Associated with spinifex or canegrass on dune-slopes and interdune flats.

Diporiphora winneckei. Windorah.

Genus *GOWIDON* Monotypic genus.

Long-nosed Dragon
Gowidon longirostris
SVL 114mm

Vertebral crest of enlarged scales continuous with moderately high nuchal crest; remaining dorsal scales small and uniform with keels converging back towards midline; exposed tympanum; long-limbs, and extremely long, slender tail; very long snout. Grey to reddish brown with prominent pale stripe along upper jaw, discontinuous or narrowly continuous with sharp-edged pale dorsolateral stripe to hips, black patch behind ear encl. a white spot, a series of short, dark dorsal bars, narrow pale midlateral stripe and, on adults, 3 prominent orange spots on each side of belly.

Far wstn CHC. Also deserts from central Aust. to west coast. Usually associated with tree-lined drainage systems and gorges, but also where trees grow in spinifex deserts.

Gowidon longirostris. Alice Springs, NT.

Genus *INTELLAGAMA* Monotypic genus.

Water Dragon
Intellagama lesueurii
SVL 245mm

Pronounced spiny nuchal and vertebral crest; laterally compressed body and tail; small uniform body scales with widely spaced transverse rows of larger scales; exposed ear; long powerful limbs. Olive green to brown with short dark bands on body, dark stripe behind eye and red flush on chest. Mature males are largest, with stronger crests, heavier jowls, brighter colours.

Margins of waterways in NET, SEQ, BB, CQC, EIU, WT, n. to about Cooktown. One record (probably waif carried by floods) inland at Warrego R., Charleville. Also estn NSW and Vic. Basks on creek or river banks and overhanging branches, readily taking to water if approached, sometimes dropping from many metres. Powerful swimmer; can remain submerged for many minutes and has even been recorded sleeping underwater. Abundant in densely settled towns and cities (incl. inner Brisbane), where lizards habituated to humans grow fat on scraps and handouts. Represented in Qld by ssp. *I. l. lesueurii*.. One other ssp. in sthn NSW, ACT, Vic.

Intellagama lesueurii. Brisbane.

Ta-ta Lizards

Genus *LOPHOGNATHUS*

2 Aust spp.; both reportedly occur in Qld. Dorsal scales with keels parallel to midline; weakly heterogeneous with erectable nuchal and vertebral crest of enlarged scales and slightly enlarged dorsolateral rows, and remaining dorsal scales smaller; exposed tympanum; pink mouth-lining; robust build with long limbs and tails. Typically brown or grey with prominent pale stripes, straight-edged or deeply notched to form elongate blotches. Broad pale stripe on upper and lower lips, not continuous with pale dorsolateral stripes, and often a well-defined narrow pale stripe between eye and ear. Mature males black with sharply contrasting white markings. Both spp. extremely similar, distinguished by presence or absence of a spot on the tympanum.

Semi-arboreal in tropical woodlands and margins of watercourses in NWH, GUP, MGD, EIU. Also nthn NT and WA.

R. Coupland

Lophognathus gilberti. Jabiru, NT.

Lophognathus horneri. Mt Isa.

Gilbert's Dragon

Lophognathus gilberti
SVL 86mm

No well-defined white spot on tympanum, though diffuse pale smear may be present, without any associated black pigmentation.

Woodlands and tropical savannahs in NWH and possibly far wstn GUP. Also nthn NT and ne. WA.

Horner's Dragon

Lophognathus horneri
SVL 86mm

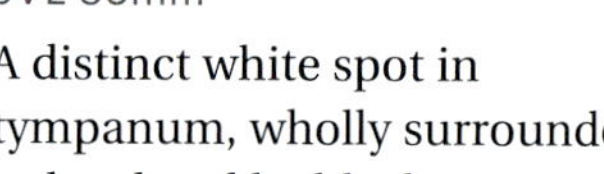

A distinct white spot in tympanum, wholly surrounded or bordered by black.

Eucalypt woodlands, tropical savannahs and vegetated margins of watercourses in NWH, GUP, MGD and EIU. A pop. in Townsville in far nthn BB is probably introduced. Also nthn NT to nw WA.

Rainforest Dragons

Genus ***LOPHOSAURUS***

2 Aust. spp.; both occur in Qld. one other in NG. Large, spiny nuchal and vertebral crests; deep dewlaps; acutely angular brows; strongly laterally compressed bodies; long slender limbs.

Restricted to widely separated se. and ne. rainforests. These secretive, slow-moving dragons perch motionless on tree trunks, from which they survey surrounding forest floor for moving invertebrates. They seldom draw attention with a noisy dash for cover, preferring to slide quietly from view if approached, keeping the trunk between themselves and potential danger. Eggs are deposited in very shallow scrapes, normally in a clearing or beside a track.

Lophosaurus boydii. Lake Barrine.

Lophosaurus spinipes. Main Range NP.

Boyd's Forest Dragon

Lophosaurus boydii

SVL 150mm

Enlarged plate-like scales on lower corner of jaw; high nuchal and vertebral crests discontinuous above level of forelimbs; large tooth-like spines along leading edge of dewlap. Rich purplish brown to greenish brown or grey with black flush on side of neck, bisected by broad cream to white bar.

Endemic. WT, between Paluma Ra. and Shipton's Flat. Lowland pops prefer to lay eggs in closed forests rather than clearings, presumably because ambient temperatures are uniformly high.

Southern Angle-headed Dragon

Lophosaurus spinipes

SVL 110mm

No enlarged plate-like scales on lower corner of jaw; high nuchal and low vertebral crest continuous above level of forelimb; limbs covered with enlarged scattered spiny scales. Brown, grey to greyish green, with or without irregular mottling, blotches or variegations.

Subtropical rainforests of SEQ, n. to Gympie area. Also nthn NSW. In early December, large numbers of females may be encountered laying eggs beside forest tracks.

Genus *MOLOCH* Monotypic genus.

Thorny Devil
Moloch horridus
SVL 110mm

Moloch horridus. Ethabuka.

Distinctive spiny hump on neck; large thorn-like spines over head, body, limbs and tail; very robust, with short limbs, short thick tail. Orange-red to yellow with large, prominent, sharp-edged darker blotches and narrow pale vertebral and dorsolateral stripes.

Sandy deserts of CHC, in vicinity of Ethabuka. Also NT, SA, WA. Spectacular and slow-moving; walks like jerky clockwork toy. Feeds exclusively on small black ants, positioning itself above a continuous trail to consume up to 5000 per meal with rapid dabs of short thick tongue. It visits regular defecation sites. Minute capillary-like channels between scales allow water taken up via direct contact with skin to flow to corners of mouth to be swallowed.

Bearded Dragons
Genus *POGONA*

6 Aust. spp.; 3 occur in Qld. Numerous spines scattered over body, limbs and tail, including long spines along dorsoventral angle of body, a row across base of head and usually a row across throat forming a 'beard'; exposed ear; relatively short limbs; dorsally depressed body.

Widespread throughout Qld except CYP, mainly in dry forests, woodlands and many urban areas, although *P. henrylawsoni* is restricted to treeless grasslands. Bearded Dragons perch conspicuously on limbs, stumps and fence posts. When approached they often freeze with head lowered, but if harassed they extend the spiny 'beard' and gape the mouth, radically altering their appearance. They have extremely broad diets, consuming fruits, flowers and soft leaves, invertebrates and small vertebrates.

KEY TO *POGONA*

1 Transverse rows of spiny scales under throat strong and prominent; more than 18 lamellae under 4th toe .. 2
Transverse rows of spiny scales under throat weak to absent; fewer than 18 lamellae under 4th toe .... ***henrylawsoni***

2 Head relatively narrow, with backward-curving row of thick spines on base (**a**); rings of enlarged scales along tail, separated by bands of smaller scales; mouth-lining yellow........ ***barbata***

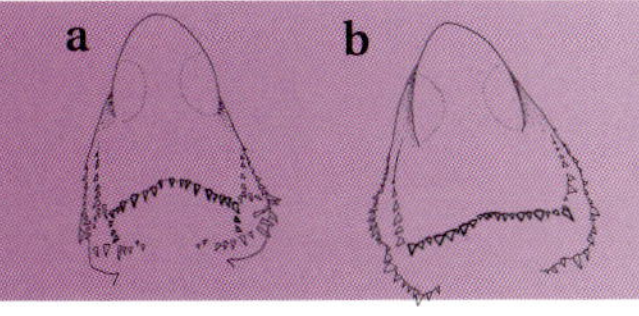

Head broad, with row of thick spines forming a nearly straight line across base (**b**); scales along tail roughly uniform, with no enlarged rings; mouth-lining pink .... ***vitticeps***

Pogona barbata. Glenmorgan.

Pogona barbata. Defensive pose. Dutton Park.

Common or Eastern Bearded Dragon

Pogona barbata

SVL 250mm

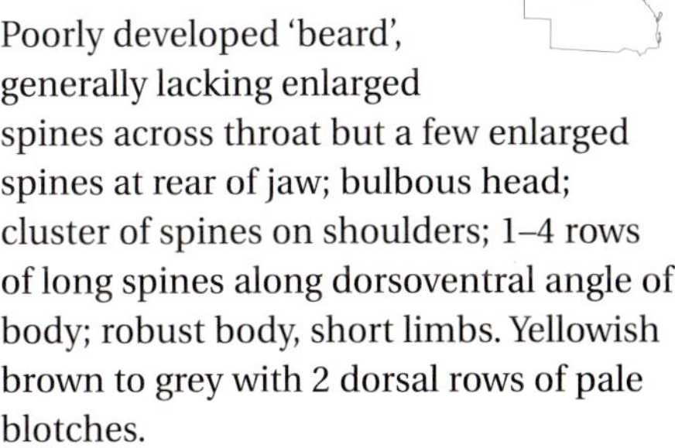

Well-developed squarish 'beard' (when erected); relatively narrow head; backward-curving arc of spines across rear of head; numerous rows of long spines along dorsoventral angle of body; rings of enlarged scales along tail, separated by bands of smaller scales; bright yellow mouth-lining. Grey with 2 dorsal rows of pale blotches. Mature males have black 'beards', particularly when displaying.

Woodlands, dry sclerophyll forests, many urban and rural areas in NET, SEQ, BB, CQC, DEU, EIU, WT, n. to Cairns area. Also NSW, Vic, SA.

Pogona henrylawsoni. Winton area.

Downs Bearded Dragon

Pogona henrylawsoni

SVL 130mm

Poorly developed 'beard', generally lacking enlarged spines across throat but a few enlarged spines at rear of jaw; bulbous head; cluster of spines on shoulders; 1–4 rows of long spines along dorsoventral angle of body; robust body, short limbs. Yellowish brown to grey with 2 dorsal rows of pale blotches.

Endemic. Restricted to treeless grassy plains of GUP, MGD, between about Croydon and Augathella. These grasslands offer few elevated perching sites but provide secure shelter in deep, labyrinthine soil cracks.

Central Bearded Dragon

Pogona vitticeps

SVL 250mm

Well-developed roundish 'beard' (when erected); broad head; straight line of spines across rear of head; only 1–2 rows of long spines along dorsoventral angle of body; mouth-lining normally pink. Grey to rich orange with 2 dorsal rows of pale blotches, sometimes bright orange flush on side of head, particularly around eye. Mature males often have sharply contrasting black 'beards'.

Semi-arid to arid woodlands, shrublands in ML, CHC, MGD, NWH. Also NSW, Vic, SA, NT.

Pogona vitticeps. Defensive pose. Nappa Merrie Stn.

Pogona vitticeps. Currawinya NP.

Genus *TROPICAGAMA* Monotypic genus.

Swamplands Lashtail

Tropicagama temporalis

SVL 120

Vertebral crest of enlarged scales continuous with moderately high nuchal crest; remaining dorsal scales small and homogeneous with keels converging back towards midline; exposed tympanum; long limbs, and extremely long, slender tail. Brown to black with prominent pale dorsolateral stripes broadly continuous with broad pale stripe through upper and lower lips. No well-defined pale stripe between eye and ear. Juv. and female often have one or more broad dark bands across back and shoulders. Breeding male very sharply marked with dark grey or black on head, chin and throat, and strongly contrasting white labial and dorsolateral stripes.

Semi-arboreal in dry tropical woodlands, usually associated with vegetation along watercourses in coastal and near coastal GUP and CYP, through Torres Strait Is. Also nthn NT and sthn NG.

Tropicagama temporalis. Saibai Is.

Earless Dragons
Genus *TYMPANOCRYPTIS*

23 Aust. spp.; 6 currently described in Qld. No visible ear (hidden beneath scaly skin); numerous enlarged spiny scales scattered over body and limbs; robust bodies; short limbs; preanal and sometimes femoral pores on males. Most are coloured to match their substrate, with patterns typically including narrow pale longitudinal stripes and broader dark bands.

Earless dragons live in dry open areas such as featureless stony deserts and sparsely vegetated cracking clay plains. Most extend across vast tracts of the interior, but some endangered grassland species have extremely restricted distributions. They are tolerant of high temperatures, and are often seen perching atop small stones or windrows on scorching summer days. Some are superb mimics, with rounded heads and bodies closely resembling the strewn pebbles amongst which they dwell. Morphological and genetic analysis has revealed at least 3 additional Qld spp. that remain undescribed. One of these, included here from central BB, appears largely dependent on croplands and is probably of high conservation concern.

KEY TO *TYMPANOCRYPTIS*

1	Scales on top of head smooth to very weakly keeled	***intima***
	Scales on top of head rough and distinctly keeled	2
2(1)	Femoral pores present	3
	Femoral pores absent	5
3(2)	5 pale stripes – vertebral, dorsolaterals and laterals; GUP	***pentalineata***
	Usually 1–3 pale stripes – vertebral and sometimes dorsolaterals; if laterals are also present, then weak	4
4(3)	3 pale spots on snout; Roma area	***wilsoni***
	Snout plain to mottled; Qld interior	***tetraporophora***
5(2)	Wavy dark and pale transverse markings between eyes; Darling Downs and mid-eastern interior	6
	Broad pale transverse band between eyes; sw CHC	***argillosa***
6(5)	Estn Darling Downs	***condaminensis***
	Clermont and Emerald areas to Collinsville	***Tympanocryptis*** sp.

Claypan Earless Dragon
Tympanocryptis argillosa
SVL 52mm

Moderately keeled scales on top of head; 2 preanal pores; no femoral pores; scattered enlarged spinose scales present, generally largest and most numerous within dark dorsal markings; remaining dorsal body scales weakly keeled to flat. Pale yellowish brown to greyish brown with broad transverse pale band between eyes, and weak pattern of pale stripes and broad dark bands. Continuous narrow vertebral stripe and broader broken dorsolateral stripes,

Tympanocryptis argillosa. Moomba area, SA.

visible only where they transect 4–5 dark bands between nape and hips.

Recorded from Buloo Downs Stn in sw CHC, in clay swales between sandy dunes. Presumably contiguous with broader distribution across Strzelecki and sthn Simpson Deserts, SA.

Tympanocryptis condaminensis. Bongeen.

Condamine Earless Dragon

Tympanocryptis condaminensis

SVL 63mm

Rough, prominently keeled scales on top of head; 2 preanal pores; no femoral pores; prominent pale stripes usually incl. lateral stripe. Brown to grey with narrow pale vertebral, dorsolateral and midlateral stripes overlaying a series of broad dark bands. Snout has 3 prominent pale spots. Lower lateral and ventral surfaces comprise strongly contrasting irregular dark and pale banding and speckling, concentrated ventrally on throat and upper chest, and with pale pigment predominating. Adults have throats flushed with bright orange (males) or yellow (females).

Endemic. Confined to small fragmented areas of estn Darling Downs, from wstn outskirts of Toowoomba w. to Dalby, n. to Pirrinuan/Jimbour area and s. to Clifton in BB. All known pops occupy road verges of remnant grassland and intensely farmed, mixed crops of cotton, maize and sorghum.

Tympanocryptis intima. Diamantina Lakes NP.

Smooth-snouted Earless Dragon

Tympanocryptis intima

SVL 61mm

Smooth scales on round head; enlarged dorsal spines arranged in longitudinal rows; 2 preanal pores on males; no femoral pores; very squat with short limbs and tail. Pale grey to brick red, usually with dark bands and indications of pale stripes.

Bleak stony plains of CHC , ML, MGD, NWH and GUP. Also adj. NSW, SA and NT. A superb pebble-mimic.

Five-lined Earless Dragon

Tympanocryptis pentalineata

SVL 60mm

Rough, prominently keeled scales on top of head; 2 preanal and 2 femoral pores; 5 prominent pale stripes. Brownish black with weak,

Tympanocryptis pentalineata. Normanton area.

Tympanocryptis sp. 'Bettafield', Gindi district.

narrow grey vertebral stripe, narrow white dorsolateral stripes, and lateral stripes comprising a row of enlarged, sharp white scales along flanks. Stripes overlay 3 dark bars. These are weak dorsally but between dorsolateral and lateral stripes they are strongly contrasting and with scattered enlarged spinose white scales.

Endemic. Currently known only from flood plains vegetated with grasses and low shrubs about 50km sw of Normanton in GUP.

Emerald Earless Dragon

Tympanocryptis sp.
SVL 60mm

Rough, prominently keeled scales on top of head; 2 preanal pores; no femoral pores; prominent pale stripes sometimes incl. lateral stripe. Brownish grey with very narrow pale vertebral stripe, broader dorsolateral stripes and sometimes a narrow pale midlateral stripe overlaying a series of broad dark bands and transverse, oval-shaped pale interspaces. Yellow speckling recorded on jowls of some females, and indistinct orange flush on gular fold, presumably of males.

Open grasslands and croplands (such as chickpea and sorghum) in Clermont and Emerald areas of BB. Pop. further ne. near Collinsville is possibly same sp.

Tympanocryptis sp. Myuna Stn, Collinsville area.

Possibly closely related to Qld's other grassland spp. Included here prior to formal description because of its probable conservation concern due to proposed coal extraction and the modification of virtually all known habitat for agriculture.

Eyrean Earless Dragon

Tympanocryptis tetraporophora
SVL 74mm

Rough, prominently keeled scales on top of head; 2 preanal and 2 femoral pores; relatively long limbs. Grey to brown or brick red with variable pattern of narrow pale vertebral and sometimes dorsolateral stripes overlaying broad dark bands. Midlateral stripe absent or, if present, usually weak and not extending forward to axilla. Mature males have yellow to orange along edges of belly and on throat; females have yellow flush on rear of belly.

Tympanocryptis tetraporophora. Morney Plain.

Vast tracts of interior, on clays, loams, stony plains with low open vegtn. Also NSW, SA, NT. In hot weather can be seen propped, dinosaur-like, on hindlimbs atop small roadside stones, bodies angled into the sun and pale bellies reflecting excess radiant heat.

Roma Earless Dragon

Tympanocryptis wilsoni

SVL 53mm

Rough, prominently keeled scales on top of head; 2 preanal and 2 femoral pores; 3 prominent pale stripes; 3 pale spots on snout. Yellowish brown with narrow pale vertebral and dorsolateral stripes overlaying a series of broad dark dorsal bands. Lower lateral and ventral surfaces with heavy brownish black speckling, concentrated ventrally on throat and upper chest, and with dark pigment predominating.

Endemic. Known only from small areas of grassland from just w. of Roma in BB to between 20 km w. and 40 km sw.

Tympanocryptis wilsoni. Mt Abundance area.

Tympanocryptis wilsoni. Mt Abundance area.

MONITORS OR GOANNAS

Freckled Monitors (*Varanus tristis*) frequently shelter in the bottle-shaped mud nests of martins.

Family **Varanidae**

Moderate-sized to enormous lizards with long snouts, sharp, recurved teeth, strong sharp claws, loose skin and slender, deeply forked tongues. Tails range from thick and spiny to strongly laterally compressed or slender and prehensile. All are placed in the genus *Varanus.*

Foraging goannas constantly protrude the tongue as they investigate burrows and crevices. Their tongue functions like that of a snake, an acutely sensitive mechanism that collects particles for chemical analysis and is capable of determining the direction of a scent source.

Goannas are predators and scavengers, with food choice largely dictated by size. The largest scavenge from carcasses and prey on mammals, birds and reptiles. Smaller goannas capture arthropods and small lizards. In each case food is swallowed with distinctive jerking motions of the head and neck.

All are egglayers. Some excavate burrows to deposit their clutches, while others rely on termite nests to serve as natural, climate-controlled incubation chambers.

The family includes the world's largest lizards. The most familiar of these in the interior and west is the Yellow-spotted Monitor (*V. panoptes*); in the forested east Lace Monitors (*V. varius*) patrol regular picnic and camping grounds in search of scraps.

KEY TO *VARANUS*

1 Distal half of tail strongly laterally-compressed 2
Distal half of tail very weakly laterally-compressed to round in cross-section 10

2(1) Nostrils directed upwards 3
Nostrils directed laterally 4

a b c d e

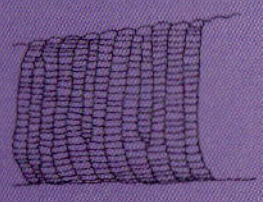

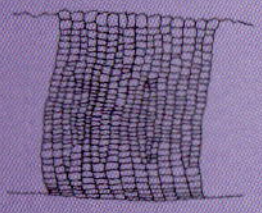

3(2) Scales around tail mostly arranged as regular rings **(a)**; throat heavily barred; moderately large dorsal keel on tail .... ***mitchelli***
Scales around tail not regular rings, as they are significantly and abruptly larger ventrally **(b)**; extremely large dorsal keel on tail .... ***mertensi***

4(2) Scales around tail mostly arranged as regular rings **(a)** 5
Scales around tail not regular rings, as they are significantly and abruptly larger ventrally **(b)** 7

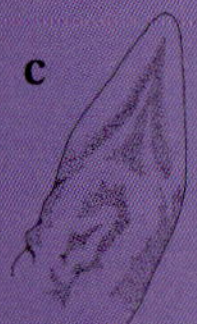

5(4) Tail scarcely longer than head and body; simple banded pattern .... ***spenceri***
Tail significantly longer than head and body; pattern incl. pale spots and pale-edged dark stripe behind eye 6

6(5) Throat marked with grey streaks or chevrons **(c)**; pale dorsal markings clustered into rosettes; tail-tip pale and patternless .... ***gouldii***
Throat marked with black spots or elongate blotches **(d)**; all pale dorsal spots, when present, sharp and round; tail-tip normally with dark bands .... ***panoptes***

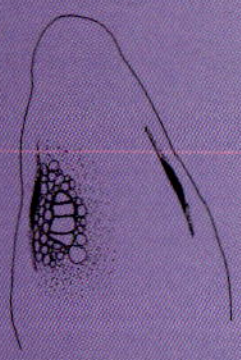

7(4) Prominent net-work of dark lines on throat .... ***giganteus***
No network of dark lines on throat 8

8(7) Pale spots, if present, arranged in bands; no enlarged supraocular scales .... ***varius***
Simple pattern of sharp, scattered pale spots; a few enlarged supraocular scales **(e)** 9

9(8) Midbody scales fewer than 145; no extensive blue pigment on tail .... ***chlorostigma***
Midbody scales more than 145; extensive blue pigment on tail .... ***doreanus***

10(1) Tail shorter than, or about as long as head and body .... ***brevicauda***
Tail noticeably longer than head and body 11

11(10) Distal half of tail striped 12
Pattern on distal half of tail transverse or absent 13

12(11) Back spotted .... ***eremius***
Back banded .... ***gilleni***

13(11) Tail very spiny 14
Tail smooth to strongly keeled but not spiny 15

14(13) Neck striped; pale ocelli on back; TL up to 63mm .... ***acanthurus***
Neck not striped; dark flecks but no pale ocelli, though they may cluster to leave a few dark-centred pale spots; TL up to 40mm .... ***storri***

15(13) Distal half of tail weakly laterally-compressed 16
Distal half of tail round in cross-section 17

f g h

16(15) Tail-tip dark .... ***semiremex***
Tail-tip cream .... ***glebopalma***

17(15) Some supraocular scales transversely enlarged (**f**); tail extremely slender and prehensile 18
No transversely enlarged supraocular scales; tail moderate, not prehensile 19

18(17) Bright green; is. of Torres Strait .... ***prasinus***
Largely black; CYP mainland .... ***keithhorni***

19(17) Sharp junction between small supraocular scales and slightly larger scales on top of head (**g**).............................. ***tristis***
Even gradation between small supraocular scales and slightly larger scales on top of head (**h**) .... ***scalaris***

Spiny-tailed Monitor
Varanus acanthurus
TL 630mm

Tail round in cross-section and very spiny; pale stripes on neck. Reddish brown with cream or yellow spots, many of them dark-centred to form small rings.

Rocky ranges and heavy to stony soils of NWH, wstn MGD, probably nthn CHC. Also NT, nthn WA. Terrestrial and rock-inhabiting, sheltering in crevices and shallow burrows under rocks and old termite mounds. Well-armoured spiny tail used to block entrance and render extraction difficult.

Short-tailed Pygmy Monitor
Varanus brevicauda
TL 230mm

Tail round to triangular in cross-section; shorter than, or about as long as, head and body; thick and very strongly keeled. Reddish brown to fawn with scattered dark flecks.

Sandy deserts of wstn CHC. Also arid zones of NT, SA, WA. Occurs only in the presence of spinifex. Terrestrial, and extremely secretive, foraging close to cover and excavating concealed burrows under bases of spinifex hummocks. World's smallest monitor lizard.

Varanus acanthurus. Lawn Hill NP.

Varanus brevicauda. Sthn NT.

Varanus chlorostigma. Lockerbie Scrub.

J. Bolton

Varanus doreanus. Iron Range.

Mangrove Monitor
Varanus chlorostigma
TL 1.2 m

Tail strongly laterally-compressed with scales not in regular rings, as they are significantly and abruptly larger ventrally; enlarged scales above eyes; fewer than 145 midbody scale rows; dark forks on tongue; simple spotted pattern. Black with small, sharp, yellow to greenish blue spots on head, body, limbs and tail. No dark bands across throat.

Coastal mangroves and rainforests on nthn CYP and Torres Strait is. Also nthn NT. Widespread from Melanesia to South East Asia. Terrestrial, arboreal and semi-aquatic, inhabiting forests and water-side vegetation. A powerful swimmer.

Blue-tailed Monitor
Varanus doreanus
TL 1.3m

Tail strongly laterally-compressed with scales not in regular rings, as they are significantly and abruptly larger ventrally; enlarged scales above eyes; more than 145 midbody scale rows; distinctly pale forks on tongue; simple spotted pattern; blue tail. Black with numerous small, sharp, yellow spots on head, body and limbs, and dark bands across throat. Tail distinctly flushed with blue.

Far nthn CYP. Also NG. Status in Aust. uncertain, and examination of Aust. specimens currently identified as *V. indicus* is warranted. Lizards conforming to *V. doreanus* recorded from rainforest at Iron Range. A related species, *V. finschi,* also reported on CYP but with no confirmed specimens or localities.

Varanus eremius. Three Ways area, NT.

Pygmy Desert Monitor
Varanus eremius
TL 460mm

Tail round to triangular in cross-section; longitudinally striped. Brick red to reddish brown with dark and pale spots on body, distinctive stripes on tail.

Sandy spinifex deserts of CHC. Also arid zones of NT, SA, WA. Terrestrial, foraging widely amongst spinifex hummocks, readily investigating any fresh diggings and leaving distinctive tracks with straight tail drag marks. Shelters in deep burrow, normally set at base of spinifex hummock.

A. Emmott

Varanus giganteus. Lark Quarry.

Perentie

Varanus giganteus

TL 2.4m

Tail laterally compressed, with scales not in regular rings, as they are significantly and abruptly larger ventrally; long neck; angular brow; black net-like lines on throat; extremely large size. Cream with heavy dark speckling and bands of large, circular, dark-edged pale spots across body and tail.

Rocky ranges, mesas of arid CHC, wstn MGD, sthn NWH. Also arid zones of SA, NT, WA. Terrestrial. Forages widely, normally sheltering in caves and deep rock crevices. Australia's largest lizard.

Varanus gilleni. Ethabuka Stn.

Pygmy Mulga Monitor

Varanus gilleni

TL 380mm

Tail round in cross-section; longitudinally striped. Grey, with narrow dark bands across body and base of tail, dark longitudinal stripes along remainder of tail.

Wstn CHC. Also arid zones of SA, NT, WA. Arboreal, sheltering under loose bark and in hollows, normally in standing trees such as mulga and desert oak. Often locally common, but secretive and infrequently encountered foraging.

G. Schmida

Varanus glebopalma. Mt Isa area.

Black-palmed Monitor

Varanus glebopalma

TL 1m

Tail very long; weakly laterally compressed; very long neck; dark net-like lines over throat and sides of head. Greyish brown with little pattern but basal three-fifths of tail black, sharply contrasting with cream tip. Palms and soles of feet black.

Rock outcrops, gorges, escarpments in NWH, south to Mt Isa area. Also nthn NT, WA. Forages with great agility over rock faces and beneath overhangs, sheltering in crevices and caves.

Sand Goanna; Gould's Goanna

Varanus gouldii

TL 1.6 m

Tail laterally-compressed, with scales mostly arranged as regular rings; diffuse grey streaks or chevrons on throat; pale-edged dark stripe extending back from eye; cream to yellow, patternless tail-tip. Brown to yellowish or reddish brown with

Varanus gouldii gouldii. Mitchell area.

Varanus gouldii flavirufus. Nappa Merrie Stn.

dense blackish brown and pale yellow speckling, the pale pigment tending to cluster into tight, often transversely arranged, rosettes or dark-centred spots. Ill-defined ssp., *V. g. flavirufus* occupies interior deserts. At up to about 1.5 kilograms, *V. gouldii* is much lighter than the similar *V. panoptes*.

Throughout Qld with possible exception of parts of MGD, DEU and BB. Habitats range from spinifex deserts of CHC to dry sclerophyll forests and coastal heaths of SEQ. Also extends across all mainland states. Favours sandier habitats than *V. panoptes*. Terrestrial, sheltering in sloping burrows, often excavating several within home range.

Canopy Goanna
Varanus keithhorni
TL 770mm

Extremely slender build; tail very long, thin, round in cross-section, prehensile; long limbs and digits; large, transversely elongate scales above eyes; sombre colouration. Very dark grey to nearly black with fine cream spots either scattered or concentrated to form thin paired chevrons on body and rings on tail. Tip of snout sometimes light bluish green.

Varanus keithhorni. Iron Ra. area.

Endemic. Rainforests, vine thickets between McIlwraith and Iron Ra., CYP. Canopy inhabitant, foraging with ease amongst slender branches and outer foliage. Other arboreal monitors tend to occupy trunks and larger branches.

Mertens' Water Monitor
Varanus mertensi
TL 1.1m

Very strongly laterally compressed tail with extremely high vertebral keel; caudal scales not in regular rings, as significantly and abruptly larger ventrally; nostrils directed upwards on upper part of snout; simple pattern of spots. Olive grey with numerous small, dark-edged pale spots over body, limbs and tail.

Margins of watercourses in CYP, EIU, GUP, NWH s. to Mt Isa area. Also nthn NT, WA. Inhabits pandanus, paperbarks,

R. Valentic

Varanus mertensi. Lake Moondarra, Mt Isa area.

Varanus panoptes. Tambo.

other thick waterside vegtn. Semi-aquatic and arboreal, readily taking to water if disturbed, and frequently when foraging. Numbers have declined sharply as cane toads have increased and spread.

Mitchell's Water Monitor

Varanus mitchelli

TL 700mm

Laterally compressed tail with moderately high vertebral keel and caudal scales in regular rings; nostrils directed upwards on upper part of snout. Dark grey with pale spots and/or dark-centred ocelli on back. Throat and sides of neck yellow with black spots and bars.

Recorded from Gregory R. in nthn NWH. Also nthn NT, nthn WA. Tropical waterways lined with pandanus, paperbarks, other thick vegtn. Skilled climber and swimmer, foraging along banks and in trees, often taking to water if pursued. Shelters in hollows and behind loose bark.

G. Schmida

Varanus mitchelli. Gregory R., Riversleigh.

Yellow-spotted Monitor

Varanus panoptes

TL 1.4m

Tail very strongly laterally compressed with scales mostly arranged in regular rings; pale-edged dark stripe extending back from eye; pattern comprising round pale spots; black-spotted throat; tail-tip often banded; large size, robust build. Blackish brown with bands of circular pale spots across back, limbs and tail, often fading on anterior body. Throat and belly marked with dark spots, usually incl. large, elongate spots along edges of throat. At up to 4kg, much heavier than similar *V. gouldii*.

Widespread across Qld, e. to wstn Brisbane suburbs, some offshore is. Also nthn NT, WA. Mainly occurs on hard soils and heavy loams; also occupies sandy soils with heaths near coast. Common very large monitor of outback, but still much confusion between this and *V. gouldii*, particularly in SEQ.

Emerald Monitor; Wyniss

Varanus prasinus

TL 750mm

Extremely slender build; tail very long, thin, round in cross-section, prehensile; long limbs and digits; large, transversely elongate scales above

Varanus prasinus. Moa Is.

Varanus scalaris. Badu Is.

Varanus scalaris. Burdekin district.

eyes. Bright green with black chevrons across back or fine dark netted pattern.

Torres Strait is., s. to Moa. Also NG. Arboreal, foraging amongst slender branches and outer foliage in upper canopy of rainforests, monsoon forests. Known as 'Wyniss' on Moa Is.

Spotted Tree Monitor

Varanus scalaris

TL 600mm

Tail round in cross-section; even gradation from larger scales on top of head to smaller scales above eyes; spotted pattern. Grey, brown to black with extremely variable pattern dominated by pale spots; usually dark-centred to form ocelli, but often solid or fragmented into pale flecks. One colour form (informally referred to, but never described as, '*V. pellewensis*') has fragmented ocelli overlaid with broad reddish brown bands.

Rainforests of WT to tropical woodlands of nthn BB, EIU, CYP, Torres Strait is., NWH, GUP, CHC. Also nthn NT, WA. Arboreal, sheltering in tree hollows and under loose bark.

Rusty Monitor

Varanus semiremex

TL 600mm

First third of tail round in cross-section, but last two-thirds weakly laterally compressed; dark, smoothly polished scales on top of head.

Varanus semiremex. Mid-eastern Qld.

Greyish brown with dark flecks, often forming net-like pattern. Throat and chest rusty yellow. Pops from far n. often darker, with dark-centred pale ocelli.

Endemic. Margins of waterways, incl. coastlines and estuaries, along e. coast from Gladstone in BB to CQC, WT, CYP. Shelters in hollow limbs and tree trunks, particularly mangroves. In areas adj. to fresh water, trees such as paperbarks are occupied.

A. Emmott

Varanus spenceri. Stonehenge.

Spencer's Monitor

Varanus spenceri

TL 1.2m

Tail laterally compressed and only about as long as head and body; very robust build; banded pattern. Grey to greyish brown with dark vertical bars on lips, simple pale bands across body and tail, tending to form V-shapes on nape.

Restricted to treeless grasslands on deeply cracking clay in MGD. Also adj. areas of NT. Terrestrial, sheltering in soil cracks.

Varanus storri storri. Charters Towers.

Vananus storri ocreatus. Adels Grove.

Storr's Monitor

Varanus storri

TL 400mm

Tail round in cross-section; very spiny; no pale stripes on neck. Brown to reddish brown, often conspicuously paler along midline, with flecks, broken ocelli or net-like pattern over body and limbs. *V. s. storri* has uniformly small scales on limbs. *V. s. ocreatus* has enlarged scales beneath hindlimbs, towards the feet.

V. s. storri is endemic in EIU, nthn DEU. *V. s. ocreatus* occupies NWH, nthn NT, WA. Terrestrial and rock-inhabiting, sheltering in narrow crevices or excavating burrows under rocks. Sometimes locally abundant where low weathered outcrops occur in dry tropical woodlands.

Varanus tristis. Sundown NP.

Varanus tristis. Mt Isa area.

Black-headed Monitor; Freckled Monitor

Varanus tristis

TL 760mm

Tail round in cross-section and strongly keeled; sharp delineation between larger scales on top of head and smaller scales above eyes. Extremely variable; pale grey to brown or reddish brown with bands of dark-centred pale ocelli, and dark patternless tail, except for narrow pale rings on base. Some pops have black patternless head and neck; on others body colour and pattern extend well onto base of head.

Throughout Qld, excl. only SEQ, most of NET. Also NT, WA, nthn SA, nw. NSW. Dry forests, woodlands, rock outcrops. Arboreal and rock-inhabiting. Often found in mud nests of swallows under culverts and rock overhangs.

Ssp. of dubious status. *V. t. tristis* applies to lizards with black patternless heads and necks. *V. t. orientalis* refers to those with pattern extending onto neck.

Lace Monitor

Varanus varius

TL 2.1m

Tail long, slender and laterally compressed, with scales not in regular rings, as significantly and abruptly larger ventrally; very broad bands on distal half of tail; chin usually banded; ridge of large scales along inner edge at base of 4th toe; very large size. Dark grey to black with numerous narrow bands of prominent pale spots or flecks. In some drier parts of range distinctive 'Bell's form' also occurs. This has very broad, simple black and yellow bands across body and tail but normally none under throat. Juv. of both colour forms intensely patterned.

Well-timbered estn areas, in NET, SEQ, BB, estn ML, w. nearly to Charleville, estn EIU, WT n. to Cairns area. Also NSW, Vic, estn SA. Arboreal, foraging widely on ground but taking to trees if disturbed. Eggs laid in termite nests where they are entombed by the termites. Becoming increasingly apparent that young require their mother's assistance to emerge.

Varanus varius. D'Aguilar NP.

Varanus varius. 'Bell's form'. Sundown NP.

BLIND SNAKES

Blind snakes' eyes, while greatly reduced, are visible beneath a protective ocular scale. *Anilios wiedii*. Glenmorgan.

Family **Typhlopidae**

Worm-like, non-venomous burrowing snakes with cylindrical bodies, bluntly rounded tails tipped with a short spur (absent on one species), close-fitting glossy scales, small ventral scales undifferentiated from adjacent scales, small eyes protected beneath head scales and small mouths set back below the snout.

These secretive snakes dwell in soil cavities, in ant and termite nests, under thick leaf litter and compost. They come to the surface infrequently, always at night and usually after rain. All are wholly insectivorous, consuming termites and the eggs, larvae and pupae of ants. Some appear to specialise on a narrow range of prey.

More than 250 spp. described world-wide; 48 in Aust; 20 in Qld. All native mainland Aust. spp. are in the genus *Anilios. Indotyphlops braminus* is introduced.

KEY TO *ANILIOS* AND *INDOTYPHLOPS*

1 Tail tipped with a short spine 2
Tail-tip smooth, with no spine .... ***A..aspina***

2(1) Nasal cleft joins 1st upper labial scale **(a)** 3
Nasal cleft joins 2nd upper labial scale **(b)** or preocular scale **(c)** 8

3(2) Midbody scales in 24 rows 4
Midbody scales in fewer than 24 rows 6

4(3) Snout long with an acute transverse cutting edge .... ***A. unguirostris***
Snout round from above and in profile 5

5(4) Body extremely thick; rostral scale narrow, about 2 times as long as broad when viewed from above .... ***A. ligatus***
Body moderately slender; rostral scale broad, less than 1.5 times as long as broad .... ***A. yirrikalae***

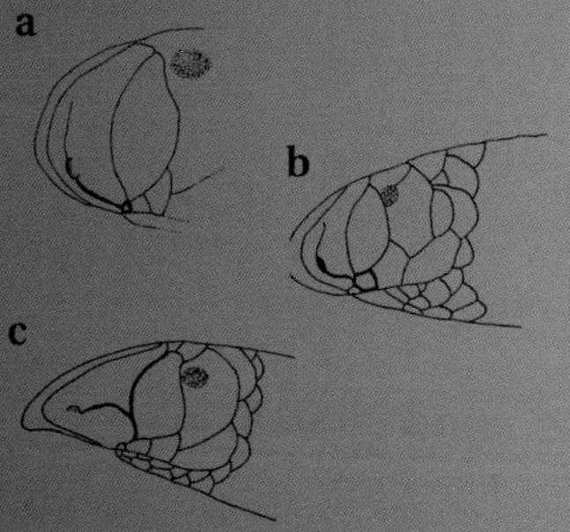

6(3) Midbody scales in 22 rows .... ***A. nigrescens***
Midbody scales in fewer than 22 rows 7

7(6) Midbody scales in 16 rows .... ***A. insperatus***
Midbody scales in 20 rows .... ***A. proximus***

8(2) Nasal cleft joins 2nd upper labial scale **(b)** 9
Nasal cleft joins preocular scale **(c)** 18

9(8) Midbody scales in 18 rows 10
Midbody scales more than 18 rows 12

10(9) Tail black 11
Tail not black .... ***A. affinis***

11(10) Body striped; snout round from above and in profile .... ***A. chamodracaena***
Body not striped; snout hooked and beak-like in profile .... ***A. grypus***

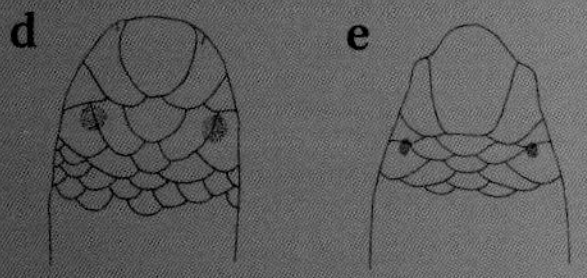

12(9) Midbody scales in 20 rows 13
Midbody scales in 22 rows 17

13(12) Snout rounded from above **(d)** 14
Snout strongly trilobed from above **(e)** .. ***A. bituberculatus***

14(13) Ventral surfaces as dark as dorsal surfaces .. ***A. leucoproctus***
Ventral surfaces paler than dorsal surfaces 15

15(14) Extremely sharp junction between dark dorsal and pale ventral colours .... ***A. silvia***
Dorsal and ventral colours more or less merge together 16

16(15) Body striped .... ***A. broomi***
Body not striped .... ***A. wiedii***

17(12) Rostral scale circular when viewed from above; extremely sharp junction between dark dorsal and pale ventral colours .... ***A. robertsi***
Rostral scale elongate when viewed from above; dorsal and ventral colours more or less merge together .... ***A. torresianus***

18(8) Midbody scales in 20 rows; snout rounded from above and in profile 19
Midbody scales in 22 rows; snout weakly trilobed from above and angular in profile .... ***A. endoterus***

19(18) Black above; ventral scales fewer than 340; nasal cleft extends well onto top of snout .... ***I. braminus***
Brown to purplish brown above; ventral scales more than 380; nasal cleft only just visible from above .... ***A. diversus***

Anilios affinis. Miriam Vale.

Small-headed Blind Snake
Anilios affinis
TL 220mm

Midbody scales in 18 rows; nasal cleft not visible from above; nasal cleft completely divides nasal scale, extending between rostral scale and 2nd upper labial scale; snout rounded from above, bluntly angular in profile; slender build. Pinkish brown above, paler below, colours merging gradually.

Dry open woodlands in sthn EIU, DEU, BB, SEQ, ML. Also nthn NSW, Solomon Is. No other vertebrate shares this distribution, suggesting separate pops may be different spp.

No-spined Blind Snake
Anilios aspina
TL 278mm

Tail smoothly rounded, with no spine on tip; midbody scales in 18 rows; nasal cleft visible from above; nasal cleft does not completely divide nasal scale, extending from above nostril to 2nd upper labial scale; snout rounded from above, slightly flattened in profile; slender build. Pink with faintly dark-edged scales.

Endemic. Known from only 2 localities; near Barcaldine and Julia Creek. Probably restricted to treeless grasslands of MGD.

E. Vanderduys

Anilios aspina. Julia Creek.

Anilios bituberculatus. Currawinya NP.

Prong-snouted Blind Snake
Anilios bituberculatus
TL 450mm

Midbody scales in 20 rows; nasal cleft not visible from above; nasal cleft does not completely divide nasal scale, extending from between nostril and rostral scale to 2nd upper labial scale; snout strongly trilobed from above, slightly angular in profile; moderate to slender build. Pink to almost black above and cream to pinkish white below, colours merging or meeting along a weakly defined junction.

Dry forests, woodlands, shrublands of CHC, ML, BB. Also dry parts of sthn Aust. from NSW to WA.

Faint-striped Blind Snake
Anilios broomi
TL 250mm

Midbody scales in 20 rows; nasal cleft not or scarcely visible

Anilios broomi. Cooktown area.

from above; nasal cleft completely divides nasal scale, extending from rostral scale to 2nd upper labial scale; snout rounded from above and in profile; small size, slender build; striped pattern. Pale brown, dark grey on tail, with 11–15 narrow, dark reddish brown stripes on dorsal and lateral surfaces. Ventral surfaces whitish and patternless.

Endemic. Restricted to belt of dry tropical woodlands of EIU from about Innot Hot Springs ne. to near Cooktown.

Cape York Striped Blind Snake

Anilios chamodracaena

TL 210mm

Midbody scales in 18 rows; nasal cleft not visible from above; nasal cleft does not completely divide nasal scale, extending from halfway between nostril and rostral scale to 2nd upper labial scale; snout rounded from above and in profile; small size, slender build; striped pattern.

E. Vanderduys

Anilios chamodracaena. Archer River.

Yellowish cream to whitish with black tail and sometimes head, prominently marked above and below with usually 18 (rarely 16) narrow dark stripes.

Endemic. Tropical woodlands of CYP, estn GUP, from Weipa s. to Inkerman Stn.

R. Valentic

Anilios diversus. Mt Isa area.

Northern Blind Snake

Anilios diversus

TL 352mm

Midbody scales in 20 rows; nasal cleft just visible from above; nasal cleft often completely divides nasal scale, extending from or from very near rostral scale to pre-ocular scale; snout rounded from above and in profile; moderately slender build. Pinkish brown to purplish brown, merging with cream to pale brown ventral surfaces.

Dry tropical woodlands in NWH, wstn GUP, se. to about Morven in wstn BB (type locality). Also nthn NT, WA.

Desert Blind Snake

Anilios endoterus

TL 376mm

Midbody scales in 22 rows; nasal cleft not visible from above; nasal cleft may or may not completely divide nasal scale, extending from rostral scale, nostril or between them to preocular

R. Valentic

Anilios endoterus. Wombah Gate area.

scale; snout weakly trilobed from above, angular in profile; moderately slender build. Pink to dark purplish brown or reddish brown, paler on snout and ventral surfaces, junction between upper and lower colours normally abrupt and ragged.

Sandy spinifex deserts of CHC. Also arid zones of all mainland States except Vic.

A. Emmott

Anilios grypus. Lochern NP.

Northern Beaked Blind Snake

Anilios grypus

TL 415mm

Midbody scales in 18 rows; nasal cleft not or scarcely visible from above; nasal cleft may or may not completely divide nasal scale, extending from rostral scale, nostril or between to 2nd upper labial scale (rarely to 1st labial or preocular); snout angular from above, hooked and beak-like in profile; extremely slender build. Pink to dark reddish brown with cream snout, black head, black tail, pale ventral surfaces, upper and lower colours merging gradually. Status of Qld pops uncertain.

Probably occupies broad range of dry, open habitats in GUP, NWH, MGD, DEU, BB, ML, CHC. Also NT, nw. WA.

Anilios insperatus. Preserved holotype. Warrill View.

Fassifern Blind Snake

Anilios insperatus

TL 96mm

Midbody scales in 16 rows; nasal cleft not visible from above, extending from 2nd upper labial scale but does not reach rostral scale; snout weakly trilobed from above and bluntly angular in profile. Very small and moderately slender. Colour in life unknown. In preservative, uniformly cream above and below with no indication of any pigment.

Endemic. Just one specimen known, from Warrill View in Fassifern Valley, SEQ. Recorded under a rock on clay soil in cleared pasture adj. to stony slope with open forest of Eucalyptus crebra, *E. moluccana* and *Corymbia citriodora.* No known live image available.

Anilios leucoproctus. Moa Is.

Cape York Blind Snake

Anilios leucoproctus
TL 250mm

Midbody scales in 20 rows; nasal cleft visible from above; nasal cleft does not completely divide nasal scale, extending from between rostral scale and nostril to 2nd upper labial scale; numerous tiny tubercles on head scales; whitish glands in sutures between head scales; snout rounded from above and in profile; small size, slender build. Dark purplish brown above and below.

Ne. CYP, ne. Torres Strait. Also Fly R. area of sthn NG.

Anilios ligatus. Croyden Stn.

Robust Blind Snake

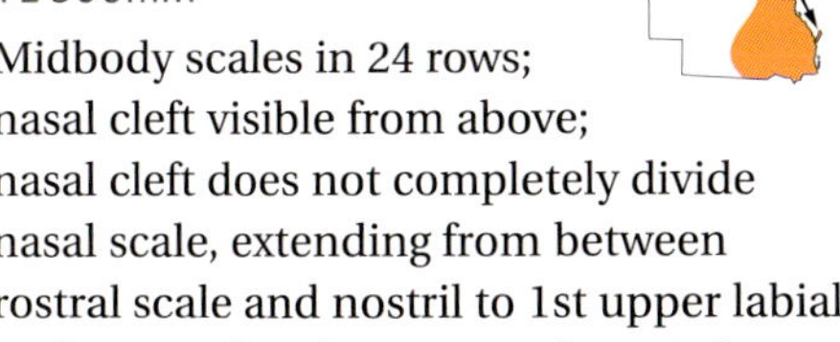

Anilios ligatus
TL 500mm

Midbody scales in 24 rows; nasal cleft visible from above; nasal cleft does not completely divide nasal scale, extending from between rostral scale and nostril to 1st upper labial scale; rostral scale narrow, about 2 times as long as broad when viewed from above; snout rounded from above and in profile; very large size, extremely thick body. Dark purplish brown to almost black above (juv. pinkish brown), cream to dull pink below, with sharp junction between upper and lower colours.

Forests, woodlands and shrublands across SEQ, BB, CQC, DEU, southern EIU, MGD and ML. Separate pops of uncertain status occur in CYP, and in NWH and GUP. Also NSW, NT, WA. Feeds on eggs, larvae, pupae of large stinging bulldog ants (*Myrmecia*), dwelling within galleries of the ants' nests, protected in part by tough scalation and perhaps by acquiring the appropriate scents to avoid triggering attack.

Anilios nigrescens. Canungra.

Blackish Blind Snake

Anilios nigrescens
TL 750mm

Midbody scales in 22 rows; nasal cleft visible from above; nasal cleft does not completely divide nasal scale, extending from near rostral scale to 1st upper labial scale; snout rounded from above and in profile; very large size and moderately robust build. Dark purplish brown to almost black above (juv. pinkish brown) with faint pale edges to each scale. Ventral surfaces pinkish white with weak ragged junction between upper and lower colours. Dark patch often present on either side of vent.

Forests, woodlands, rainforests of SEQ, NET. Also NSW, Vic.

Anilios proximus. Brisbane area.

Woodland Blind Snake

Anilios proximus

TL 750mm

Midbody scales in 20 rows; nasal cleft visible from above; nasal cleft does not completely divide nasal scale, extending from between rostral scale and nostril to 1st upper labial scale; snout bluntly trilobed from above, weakly angular in profile; very large size, robust build. Pinkish brown to grey, often with small dark patch either side of vent. Ventral surfaces pale yellowish brown to cream, upper and lower colours merging evenly.

Various well-timbered habitats in SEQ, NET, BB; outlying pops along junction of EIU and WT (Herberton, Ravenshoe and Malanda areas), and on CYP. Also NSW to Vic. One of group of extremely robust blind snakes that appear to thrive on the eggs, larvae and pupae of large stinging ants.

Roberts' Blind Snake

Anilios robertsi

TL 290mm

Midbody scales in 22 rows; nasal cleft just visible from above; nasal cleft does not completely divide nasal scale, extending from near rostral scale to 2nd upper labial scale; snout rounded from above and in profile. Very dark purplish brown above with pale snout and cream below, junction between upper and lower colours very sharply delineated and straight.

Endemic. Only one specimen collected in WT, at Romeo Ck near Shipton's Flat, s. of Cooktown. Recorded emerging from hollow fallen timber and feeding on ants' eggs in open eucalypt forest.

L. Roberts

Anilios robertsi. Romeo Ck, Shipton's Flat area.

Anilios silvia. Palmview.

Cooloola Blind Snake

Anilios silvia

TL 175mm

Midbody scales in 20 rows; nasal cleft visible from above; nasal cleft does not completely divide nasal scale, extending from near rostral scale to 2nd upper labial scale; snout rounded from above and in profile; very small size and slender build. Shiny black above, white below, with jagged, very sharply delineated junction between upper and lower colours. In some specimens, lateral edges of scales appear pale, creating 11 broad black stripes.

Endemic. Confined to narrow band of white coastal sands between Fraser Is. and Palmwoods, SEQ. Occupies rainforests, woodlands, heaths, sheltering in sand under logs and leaf litter. This tiny thread-like sp. may be Aust.'s smallest native snake.

Anilios torresianus. Cardwell.

North-eastern Blind Snake
Anilios torresianus
TL 400mm

Midbody scales in 22 rows; nasal cleft visible from above; nasal cleft does not completely divide nasal scale, extending from near rostral scale to 2nd upper labial scale, or suture between 1st and 2nd; snout rounded from above and in profile; large size, moderately robust build. Greyish brown above (juv. pinkish brown), paler on snout and tail-tip, with pale-edged scales forming obscure reticulum. Paler below, upper and lower colours merging evenly.

Variety of well-timbered habitats in CQC, nthn BB, WT, estn EIU and CYP. Also NG.

Claw-snouted Blind Snake
Anilios unguirostris
TL 490mm

Midbody scales in 24 rows; nasal cleft barely or not visible from above; nasal cleft may or may not completely divide nasal scale, extending from rostral scale, or between rostral and nostril, to 1st upper labial scale; snout long and weakly trilobed from above, hooked with prominent transverse cutting edge in profile. Dark olive brown above, cream below, colours in moderately sharp contrast.

Woodlands of sthn CYP, EIU, DEU, BB. Also across nthn Aust. to WA.

Anilios unguirostris. Moranbah area.

Anilios wiedii. Glenmorgan.

Brown-snouted Blind Snake
Anilios wiedii
TL 300mm

Midbody scales in 20 rows; nasal cleft visible from above; nasal cleft may or may not completely divide nasal scale, occasionally extending from rostral scale, but usually from between rostral and nostril, to 2nd upper labial scale; snout rounded from above and in profile; slender build. Pink to pinkish brown above, usually with dark streak on rostral scale, upper colour merging gradually with cream ventral surfaces.

Woodlands, other open habitats in SEQ, BB. Also nthn NSW.

G. Shea

Anilios yirrikalae. Preserved holotype. Yirrkala Mission, NT.

Yirrkala Blind Snake

Anilios yirrikalae

TL 200mm

Midbody scales in 24 rows; nasal cleft does not divide nasal scale, extending from 1st upper labial scale to just beyond nostril; rostral scale broad, less than 1.5 times as long as wide; snout rounded from above and in profile. Small and moderately slender. Brown above and whitish below.

Far nw. NWH, adj. to NT border. Also ne. NT. No known live image available.

Flowerpot Snake

Indotyphlops braminus

TL 170mm

Midbody scales in 20 rows; nasal cleft easily visible from above; nasal cleft completely divides nasal scale, extending from rostral to preocular scale; numerous tiny tubercles on head scales; whitish glands in sutures between head scales; snout rounded from above and in profile; extremely small size, slender build. Dark purplish brown to almost black above, paler brown below with pale snout, white around vent and tail-tip.

Introduced to is. of Torres Strait, Townsville, Brisbane and probably towns in between. Also towns and settlements across nthn Aust. World's most widespread land snake, transported as stowaways amongst soil and cargo from South-East Asia (point of origin) to Pacific, Central America, Africa and is. of Indian Ocean. Successful dispersal is due in part to parthenogenesis, all-female pops requiring only one colonist to become established.

Indotyphlops braminus. Townsville.

E. Vanderduys

Indotyphlops braminus. Distinctive glands on head. Townsville.

PYTHONS

Like all pythons, Carpet Pythons (*Morelia spilota*) guard their eggs until they hatch. Burpengary.

Family **Pythonidae**

Moderate-sized to enormous non-venomous snakes with long recurved teeth and muscular bodies. They have minute vestigial hips attached to rudimentary hindlimbs known as cloacal spurs.

Pythons are constrictors, killing by enveloping their prey within powerful, ever-tightening coils. Most feed on large bulky animals; adaptations for this include a very loosely articulated skull and a high number of small body scales, allowing greater elasticity of the skin. Excepting the 2 distinctive *Aspidites* species, all pythons are equipped with infrared heat-sensory pits along the lower and sometimes upper lips, acute to variations of as little as 1/30˚C.

Pythons are mainly nocturnal, but will often move and bask by day. They are frequently found on the ground, but most are skilled climbers, foraging with equal frequency in trees and on rock faces. *Aspidites* are largely terrestrial, while the Green Python (*Morelia viridis*) is largely arboreal. Egglayers, exhibiting maternal care unique amongst Aust. snakes, coiling around eggs to guard them until they hatch.

About 38 spp. live in the Old World from Africa through Asia to Aust.; 15 spp. in Aust., 10 occur in Qld.

KEY TO GENERA

a

b

1	Pits present in at least some labial scales; subcaudal scales divided 2
	No pits present; subcaudal scales mostly single .... ***Aspidites***
2(1)	No pits in rostral scale 3
	Pits present in rostral scale 4
3(2)	Usually only one large loreal scale **(a)**; body patternless.. ***Liasis***
	At least 3 (often more) small loreal scales **(b)**; usually at least some indication of dark dorsal blotches.......... ***Antaresia***
4(2)	Dorsal head scales large and arranged as symmetrical plates .... ***Simalia***
	Dorsal head scales mostly small and fragmented .... ***Morelia***

Genus *ANTARESIA*

4 Aust. spp.; 3 occur in Qld. Head scales arranged as large symmetrical plates; heat-sensory pits present in some lower labial scales only (none in rostral); 3 or more loreal scales; divided subcaudal scales; small size (usually 1m or less). Pattern consists of solid dark blotches or broken bands against paler background.

Throughout Qld, mainly in dry, well-drained areas such as rock outcrops, but ranging from vine thickets and rainforest edges to open arid shrublands. Also dry parts of all mainland States except Vic. Mammals and birds are favoured prey, but reptiles such as large lizards are also taken. *A. maculosa* has become famous for its habit of aggregating to capture bats emerging from clefts at Mt Etna, n. of Rockhampton. Nocturnal. Often referred to collectively as Children's Pythons, after the first named species.

KEY TO *ANTARESIA*

1	Pattern variable, from absent to present; if present, prominent to obscure dorsal and lateral blotches; widespread in Qld 2
	Pattern present, represented by flecks and scattered dots; far nthn CYP and Torres Strait .... ***papuensis***
2	Dorsal blotches ragged-edged, normally prominent, and sometimes tending to coalesce into longitudinal streaks on forebody and tail; no pale lower lateral line .... ***maculosa***
	Pattern absent to prominent; when present, dark blotches smooth-edged and sometimes arranged as transversely elongate bars with pale lower lateral line present and obvious, at least anteriorly... ***childreni***

Children's Python
Antaresia childreni
TL 1.02m

Pattern variable; tending to be reduced to virtually absent in nw. monsoonal tropics and blotched through interior and e. Plain form is shades of brown, with or without weak, roughly circular blotches. Patterned form is pale brown, yellowish brown to cream, with smooth-edged, often prominent dark blotches, sometimes transversely oriented, and a conspicuous pale lower lateral line.

Antaresia childreni. Plain form. Lawn Hill NP.

Antaresia maculosa. Moranbah area.

Antaresia childreni. Patterned form. Morven area.

Plain form occupies outcrops and escarpments of NWH and wstn GUP. Patterned form (until recently referred to as Stimson's Python) is widespread through arid to semi-arid areas to monsoonal tropics, occupying woodlands, shrublands and outcrops in CHC, MGD, GUP, DEU, EIU, CYP and ML.

Spotted Python

Antaresia maculosa

TL 1.30m

Dark blotches are ragged-edged, tending to coalesce into longitudinal vertebral streaks anteriorly and posteriorly. Cream to yellowish brown with dark brown to purplish brown blotches, usually prominent. Pale ventrolateral stripe absent or poorly developed. 2 ssp. described; *A. m. maculosa* in sthn portion of range, n. to Paluma Ra., and *A. m. peninsularis* from s. of Cairns onto estn CYP.

Widespread through estn CYP, EIU, WT, DEU, BB, CQC, SEQ, in wide range of habitats from highly disturbed, cleared pastoral land near Clermont to rainforest edges near Mt Glorious. Also nthn NSW.

Papuan Spotted Python

Antaresia papuensis

TL 1.1m

Pale to dark brown with weakly contrasting pattern of small scattered spots and flecks. There is no pale lower lateral line on anterior body.

Islands of Torres Strait and NG. Specimens from nthn CYP look similar and may be referrable to *A. papuensis* but genetic assessment needed.

M. O'Shea

Antaresia papuensis. Weam, Western Province, PNG.

Genus *ASPIDITES*

2 Aust. spp.; both occur in Qld. No heat-sensory pits; large head shields arranged as symmetrical plates; subcaudal scales mostly single; narrow head, not obviously distinct from neck; simple banded pattern.

Throughout Qld, excluding SEQ and moist parts of e. Also arid interior to dry tropical areas of all States except Vic. Terrestrial pythons which shelter in rock crevices, soil cracks, abandoned burrows and hollow logs, differing from other pythons in feeding mainly on reptiles, particularly large lizards such as dragons and goannas. This is cited as a likely reason for the absence of heat-sensory pits, as all other pythons possess the pits and prefer prey that generates an internal heat source (warm-blooded mammals and birds).

Aspidites melanocephalus. Moranbah.

Aspidites ramsayi. Charleville area.

Black-headed Python

Aspidites melanocephalus

TL 2.6m

Prominent black head, neck and throat; more than 310 ventral scales. Brown to reddish brown with numerous darker bands. Pattern normally conspicuous at all ages but brightest on juv.

Variety of dry, open, wooded to rocky habitats across Qld, just penetrating SEQ near Kroombit Tops. Also nthn NT, WA. While large numbers of these slow-moving snakes are killed each year on inland highways, their frequent presence in seemingly inhospitible cleared pastoral land is testament to their ability to persist in degraded conditions.

Woma

Aspidites ramsayi

TL 2.3m

Brown to yellow head and neck; fewer than 305 ventral scales. Pale brown to olive with numerous darker bands. Juv. and most adults have prominent dark patch above each eye, though very large adults can be drab and virtually patternless.

Dry open habitats, from spinifex deserts to brigalow communities, in CHC, ML, BB e. to about Condamine district. Also nw. NSW to WA. Appears uncommon and patchy at estn limits within sthn BB, probably in part because of severe habitat loss.

Genus *LIASIS*

2 Aust. spp.; both occur in Qld. Large head shields arranged as symmetrical plates; heat-sensory pits present on some lower labial scales and on 1–2 anterior upper labials but none on rostral; one (rarely 2) loreal scale; subcaudal scales divided; very large size; no pattern.

Restricted to tropical north. Also nthn NT, WA, NG to estn Indonesia. Mainly terrestrial, though both can often be found in trees and one enters water with enough frequency to be considered semi-aquatic. Adults feed mainly on mammals; juveniles include many lizards in their diet.

Liasis fuscus. Iron Range.

Liasis olivaceus. Lawn Hill NP.

Water Python

Liasis fuscus

TL 2.5m

Iridescent glossy sheen; midbody scales in 48 or fewer rows. Dark olive brown to almost black with yellow lower flanks and ventral surfaces.

Well-vegetated tropical river and swamp margins in GUP, CYP, EIU, WT, far nthn BB, CQC, se. to about St Lawrence. Also NT, WA, NG, Indonesia. Most aquatic Aust. python. Eats mostly mammals, also waterfowl, sometimes even young crocodiles.

Represented in Aust. and NG by ssp. *L. fuscus.* One other ssp. occurs in estn Indonesia. Relationship between these forms far from clear; current taxonomic arrangement tentative at best.

Olive Python

Liasis olivaceus

TL 2.5m

Dull milky sheen; midbody scales in 58 or more rows. Brown to olive brown with white to cream ventral surfaces.

Usually associated with rocky areas, particularly caves and overhanging ledges near watercourses in NWH, GUP, MGD, se. to about Bladensburg NP. Also nthn NT, WA. Occasionally reported to aggregate in suitable shelter sites. Adults feed on mammals as large as rock wallabies.

Represented in Qld, NT, nthn WA by ssp. *L. o. olivaceus.* One other ssp. in midwstn WA.

Genus *MORELIA*

5 spp. in Aust; 2 in Qld. Head scales small, irregular and fragmented; heat sensory pits present on rostral scale, some lower labials and first few upper labials; subcaudal scales divided.

Estn, nthn and sw. Qld, and patchily distributed in all mainland States. While mainly nocturnal, all can be encountered by day, either foraging or coiled quietly in a sunny glade. They are skilled climbers (*M. viridis is* mainly arboreal) but all can often be found at ground level. *M. spilota* regularly enters buildings, residing in the rafters and preying on possums, rats and birds.

A. Emmott

Morelia spilota metcalfei. Noonbah Stn.

Carpet Python

Morelia spilota

TL 2.5–3.6m

Usually prominent blotches, bands and/or stripes; often narrow, pale ventrolateral stripe. Extremely variable colour and pattern, both within and between pops. Several poorly defined ssp. (possibly mere colour variants) have been described, 3 of which occur in Qld. *M. s. mcdowelli* (Eastern) is typically brown to olive with large, dark-edged paler blotches but little or no pattern on top of head. *M. s. cheynei* (Jungle) is normally black to very dark brown with very prominent yellow bands, blotches or stripes, incl. on top of head. *M. s. metcalfei* (Murray/Darling) is grey to reddish brown with stripes on top of head and circular, often paired, dark-edged pale grey blotches.

Widespread through WT, EIU, BB, CQC, SEQ, NET, ML, CHC; scarce records in nthn and estn CYP. Eastern form extends along e. coast and ranges, Jungle form in WT rainforests, Murray/Darling form along inland drainage systems. Ranges of Eastern and Jungle forms abut and many specimens cannot easily be allocated

Morelia spilota mcdowelli. Mt Glorious.

Morelia spilota cheynei. Tully area.

to either form. Murray/Darling Carpet Python may be geographically isolated. Various ssp. occur in all mainland States, NG.

Amongst Qld's most familiar snakes, thanks to their large size and thriving pops in many large towns and cities. In some of the older, well-vegetated inner Brisbane suburbs up to 50 per cent of dwellings show evidence of occupation by resident pythons (normally shed skins and droppings in the ceilings).

Morelia viridis. Juvenile. Iron Ra. area.

Morelia viridis. Adult. Iron Ra. area.

Green Python
Morelia viridis
TL 1.5m

Very vivid colours. Adults bright green, with bead-like vertebral row of white scales. Juv. bright yellow with purplish brown streak through eye and lines or blotches on body.

Rainforests of Iron and McIlwraith Ra., ne. CYP. Also NG. Qld's isolated pop. reflects past eras when sea levels were lower and a forested land bridge extended over what is now Torres Strait. Qld's most arboreal python. By day rests on branch or vine with body looped in distinctive coil.

Genus *SIMALIA* 6 spp. from NG to estn Indonesia: 1 Aust sp. occurs in Qld.

Scrub Python; Amethyst Python
Simalia kinghorni
TL up to 5m; reports of 8m unconfirmed

Head scales enlarged, symmetrically arranged plates; heat sensory pits present on rostral scale, lower labial scales and first few upper labial scales; relatively slender build with large, long head; extremely large size.

Simalia kinghorni. Cairns.

Simalia kinghorni. Iron Range.

Shades of brown with irregular, angular dark blotches, sometimes forming bands and streaks and an iridescent, pearly sheen. Largest sizes and brightest patterns seen in WT area, while those on Torres Strait Is. are smaller and more slender with more muted pattern.

Usually associated with rainforests, monsoon forests and vine thickets but sometimes found in nearby dry sclerophyll forests and woodlands in WT, estn EIU and CYP. Also NG and estn Indonesia. Relationships with pops outside Aust. uncertain. Largest snake in Aust. Consumes mammals as large as wallabies and tree kangaroos.

FILE SNAKES

File snakes' tongues are more deeply forked than those of other snakes. *Acrochordus arafurae*. Lawn Hill NP.

Family **Acrochordidae**

Fully aquatic, non-venomous snakes with very robust, laterally compressed bodies, loose baggy skin, minute rough scales, small ventral scales no larger than adj. scales, tiny granular head scales with no large symmetrical shields, blunt snouts with forward-placed valvular nostrils and small protrusive eyes. All are in the genus *Acrochordus*.

They are livebearers restricted to the tropics; one in fresh water, the other in coastal mangroves and estuaries. If stranded on land they are baggy, ungainly and virtually helpless, but under water they swim gracefully. They are mainly nocturnal, sheltering by day under overhanging banks or amongst submerged roots and other vegetation. They feed on fish, including relatively large prey, captured by ambush and swallowed with a haste that belies their otherwise sluggish nature.

3 spp. worldwide, extending from South-East Asia to Solomon Is.; 2 occur in Aust.; both in Qld.

Acrochordus arafurae. Lawn Hill NP.

S. Macdonald

Acrochordus granulatus. Weipa.

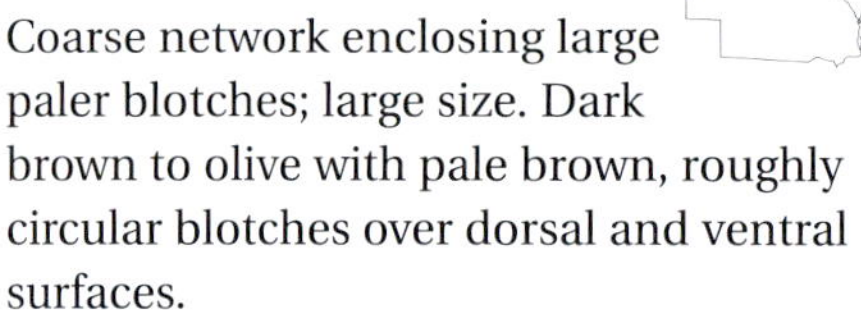

Arafura File Snake

Acrochordus arafurae

TL 2.1m

Coarse network enclosing large paler blotches; large size. Dark brown to olive with pale brown, roughly circular blotches over dorsal and ventral surfaces.

Freshwater lagoons and rivers, often lined with pandanus, in CYP, NWH, GUP. Also nthn NT, sthn NG. Introduced pop. in artificial Centenary Lake in Cairns, WT.

Little File Snake

Acrochordus granulatus

TL 1.62m

Banded pattern; relatively small size. Dark grey to brown with paler bands usually extending onto ventral surfaces and sometimes disrupted on midline.

Mangrove-lined coast, estuaries of WT, CYP, probably GUP. Also nthn Australia and from Solomon Is. to South Asia. Largely confined to salt or brackish water though occasionally found in near-coastal fresh water.

COLUBRID SNAKES

Keelbacks (*Tropidonophis mairii*) are amongst the few native Australian predators able to successfully feed on the introduced and highly poisonous Cane Toad (*Rhinella marina*). Kurwongbah.

Family **Colubridae**

This extremely diverse and complex family, containing about 1,500 species in nearly 300 genera, dominates the snake fauna of all continents except Australia. Only 5 native species occur here. All are found in Queensland, and all extend at least to New Guinea and some into South-East Asia. Of concern is an exotic terrarium escapee, the Corn Snake (*Pantherophis guttatus*), that appears to be establishing.

Colubrids are probably relatively recent arrivals. They appear to fill niches that other more long-term residents have left largely unoccupied. They include slender tree snakes and inhabitants of creek edges and other moist habitats. Colubrids feed mainly on vertebrates, with the Keelback (*Tropidonophis mairii*) preying on frogs; diurnal tree snakes (*Dendrelaphis*) hunting skinks and frogs; and the nocturnal Brown Tree Snake (*Boiga iregularis*) eating small mammals and nestling or roosting birds. *Stegonotus* includes reptile eggs on its diet.

All Australian colubrids are egglayers. Most are completely non-venomous but *B. irregularis* is a weakly-venomous, rear-fanged species. It is not regarded as dangerous but bites from any venomous snakes should be treated seriously. Colubrids can be distinguished from terrestrial elapids in possessing a loreal scale, and from mangrove and freshwater snakes in lacking valvular nostrils.

KEY TO GENERA

1	Midbody scales 27 or more	***Pantherophis***
	Midbody scales 23 or fewer	2
2(1)	Scales smooth	3
	Scales prominently keeled	***Tropidonophis***
3(2)	Anal scale divided	***Dendrelaphis***
	Anal scale single	4
4(3)	Very large eyes with vertical pupils; vertebral scale row enlarged	***Boiga***
	Small dark eyes; vertebral scale row not enlarged	***Stegonotus***

Genus *BOIGA*

About 36 spp. from Asia to wstn Pacific; 1 Aust. sp., occurs in Qld.

Brown Tree Snake; 'Night Tiger'

Boiga irregularis

TL 2m

Large bulbous head, distinct from very narrow neck; huge eyes with conspicuous vertical pupils; 19–23 midbody scale rows incl. enlarged, hexagon-shaped vertebral row; loreal scale; single anal and divided subcaudal scales. Brown to reddish brown with irregular, narrow, dark bands and cream to salmon belly in e.; cream above and below with prominent rich red bands in nw.

Widespread, excl. only large portions of ML, MGD, CHC. Occupies woodlands to rainforests and urban areas in e. and n., penetrating inland via gorges and escarpments. Also NSW, NT, WA, wstn Pacific to estn Indonesia. Skilled climber of trees, rock faces and buildings. Feeds mainly on birds and small mammals, frequently entering aviaries and hanging bird cages. Nocturnal; egglayer. Weakly venomous and rear-fanged but not normally considered dangerous.

Boiga irregularis. Mt Glorious.

Tree Snakes
Genus *DENDRELAPHIS*

About 57 spp. from wstn Pacific to South-East Asia; 2 Aust. spp., both occur in Qld. Large eyes with round pupils; 13 (occasionally 11, rarely 15) midbody scales in acutely oblique rows, including enlarged vertebral series; loreal scale; divided anal and subcaudal scales; angular ventrolateral keel along each side of body; very slender build with long thin tail.

Thickly vegetated moist parts of n. and e. Qld. Also estn NSW, nthn NT, and WA. Extremely alert diurnal snakes, strongly visually cued, raising head high and swaying gently to gauge distance. Swift, graceful climbers. Non-venomous predators of skinks and frogs. Egglayers.

Dendrelaphis calligastra. Iron Range.

Dendrelaphis punctulatus. Widespread yellow northern form. Lawn Hill NP.

Northern Tree Snake
Dendrelaphis calligastra
TL 1m

Very large eyes; black streak through eye; extremely slender build. Rich brown to olive with prominent dark streak from snout to forebody, white to yellow upper lip. Ventral surfaces often finely dark-speckled.

Rainforests, vine thickets, other areas of lush vegtn in WT, estn CYP, through Torres Strait to NG and Solomon Is. One of swiftest, most alert snakes in far n. Qld.

Common Tree Snake; Green Tree Snake
Dendrelaphis punctulatus
TL 1.2m

Moderately large eyes; no dark streak through eye; moderately slender build. Extremely variable, ranging from brown to olive green with yellow belly (common in SEQ), to black with yellow belly (common in mid e. and ne.) or yellow with pale bluish grey head and cream belly (nw. and across nthn Aust.). Occasional individuals sky blue.

Woodlands, thickly vegetated river banks, rainforests, urban areas in humid

Dendrelaphis punctulatus. Dark eastern form. Mt Glorious.

Dendrelaphis punctulatus. Typical eastern form. Kurwongbah.

parts of SEQ, NET, BB, CQC, EIU, WT, CYP, GUP, nthn NWH, and through Torres Strait to NG. Also NSW, NT, WA. Commonly seen in gardens, probably attracted to ponds where frogs breed. When disturbed, concealed pale blue edges to the scales can be suddenly revealed by distending neck and forebody.

Genus *PANTHEROPHIS*

9 spp. in North America; 1 sp. introduced to Aust.

Corn Snake

Pantherophis guttatus

TL 1.8m

Large size; midbody scales smooth to weakly keeled in 27–29 rows; anal scale divided; subcaudals mostly divided; extremely variable colour and pattern. Typical specimens orange to grey with prominent black-edged red, brown to grey dorsal blotches and smaller lateral series, ventral pattern of black chequers, and stripe under tail. However, Aust. specimens originate from illegal captive stock, much of it subjected to selective breeding for unusual colour variants.

Recorded from Sunshine and Gold Coasts. Introduced from illegal pet trade. Numerous captures incl. juvs suggest pops are becoming established. Of concern as potential competitors and carriers of pathogens.

S. Eipper

Pantherophis guttatus. Nerang.

Genus *STEGONOTUS*

About 24 spp. from NG to South-East Asia; 1 Aust. sp., occurs in Qld.

S. Eipper

Stegonotus australis. Malanda.

Slaty-grey Snake
Stegonotus australis
TL 1.3m

Squarish snout; weak ventrolateral keel; moderately small dark eyes; smooth, very glossy scales in 17 (rarely 19) midbody rows; divided subcaudal scales, single anal scale; moderately slender build; dark, patternless colouration. Very dark brown to grey above, paler on flanks and white to cream below, with iridescent sheen.

WT, s. to CQC, and n. through CYP to Torres Strait and NG. Also NT. Mainly associated with riverine habitats and moist gullies with rainforest. Largely nocturnal, foraging on the ground or in low vegetation for mammals, frogs, reptiles and their eggs. Very pugnacious if harassed, but non-venomous.

Genus *TROPIDONOPHIS*

About 19 spp. from NG to South-East Asia; 1 Aust. sp., occurs in Qld.

Keelback; Freshwater Snake

Tropidonophis mairii

TL 930mm

Moderately large eyes with round pupils; strongly keeled, matt-textured scales in 15 midbody rows; divided anal and subcaudal scales. Very variable, but typically grey to brown or olive with a mix of dark blotches or broken bands, pale flecks and often rusty lower lateral flush. Ventral surfaces cream with dark edges to scales. Upper lips pale with fine but distinctive dark sutures between scales.

Well-watered habitats near creeks or in low-lying areas, incl. disturbed urban sites in SEQ, BB, CQC, EIU, WT, CYP, GUP, nthn NWH, penetrating inland via watercourses. Also nthn Aust., ne. NSW, NG. Feeds on frogs and lizards; even enters water for tadpoles and fish. Often preys successfully on small cane toads (*Rhinella marina*), its tolerance to their poison probably inherited from related overseas spp. and genera that commonly eat toads. Diurnal and nocturnal. Non-venomous. Egglayer.

Tropidonophis mairii. Wynnum.

FRESHWATER AND MANGROVE SNAKES

Until recently, homalopsid snakes were classified as a subfamily of Colubrids. They differ in having dorsally placed valvular nostrils. Macleay's Water Snake (*Pseudoferania polylepis*) occurs in tropical freshwater streams and rivers. Claudie River.

Family **Homalopsidae**

This family of aquatic to semi-aquatic snakes contains about 40 described species occurring between India and New Guinea with diversity highest in Southeast Asia. Just five species occur in Australia, with four in Queensland.

Homalopsids are robust snakes with upward-directed eyes and dorsally placed valvular nostrils. All are weakly venomous and rear-fanged but none are regarded as dangerous. Unlike the true sea snakes, homalopsids have well-developed, broad ventral scales, and they have a round tail rather than a laterally flattened, paddle-shaped tail.

Most are confined to tropical coastal and estuarine mangrove swamps and mud flats, but Macleay's Water Snake (*Pseudoferania polylepis*) inhabits freshwater rivers and streams. Marine species favour shallow water and mud, and their shifting habitat is probably the rising and falling tides.

Macleay's Water Snake eats fish, frogs and crustaceans while marine species take fish such as mudskippers and other gobies. The White-bellied Mangrove Snake (*Fordonia leucobalia*) prefers crustaceans, frequently dismembering its prey rather than swallowing it whole. Within Australia a diet dominated by invertebrates is only shared with blind snakes (Typhlopidae). All homalopsids are live-bearers.

KEY TO GENERA

1	No loreal scale **(a)**	***Fordonia***
	Loreal scale present **(b)**	2
2(1)	Nasal scales not in contact	***Myron***
	Nasal scales in contact	3
3(2)	Scales smooth	***Pseudoferania***
	Scales prominently keeled	***Cerberus***

Genus *CERBERUS* 2 spp. from NG to India; 1 Aust. sp., occurs in Qld.

Cerberus australis. Red colour form. Darwin, NT.

Cerberus australis. Grey colour form. Buffalo Ck, NT.

Australian Bockadam

Cerberus australis

737mm

Protrusive eyes; keeled scales in 23–25 midbody rows; a loreal scale; divided anal and subcaudal scales; robust body. Grey to brick red with narrow dark bands.

Coastal mudflats with mangroves on wstn CYP, and probably across Gulf of Carpentaria. Also nthn NT, WA. Semi-aquatic, sheltering amongst mangrove roots, foraging in shallow water and exposed mud. Feeds on crabs and fish. Very pugnacious if harassed.

Genus *FORDONIA* Monotypic genus.

White-bellied Mangrove Snake

Fordonia leucobalia

TL 930mm

Broad head with short, rounded snout; no loreal scale; smooth, glossy scales in 23–29 midbody rows; divided anal and subcaudal scales (sometimes a few single subcaudals); robust build. Extremely variable in colour, from dark grey to brown, cream or pink,

Fordonia leucobalia. Wyndham, WA.

with or without contrasting blotches, spots or bands.

Recorded from mangrove-lined mudflats on nthn Torres Strait is. and estn GUP. Probably more or less continuous from wstn CYP across Gulf of Carpentaria. Also nthn NT to nw. WA, NG to South-East Asia. Mainly nocturnal, sheltering by day in crab burrows and amongst mangrove roots. Unusual dietary specialist, preying solely on crustaceans, mainly crabs. Unique amongst snakes in dismembering its prey, twisting off legs and claws before consuming body.

Genus *MYRON* 2 Aust. spp.; one occurs in Qld; one other in estn Indonesia.

Richardson's Mangrove Snake

Myron richardsonii

TL 436mm

Small, slightly protrusive eyes; a loreal scale; separated nasal scales; weakly keeled scales in 19–21 midbody rows; divided anal and subcaudal scales; long, narrow head. Yellowish brown, brick red to grey with numerous narrow irregular dark bands.

G. Stephenson

Myron richardsonii. Buffalo Creek, NT.

Recorded from coastal GUP and nthn Torres Strait. Also nthn NT, nthn WA, NG. Mangrove-lined estuaries, tidal creeks, sheltering amongst mangrove roots and in crab burrows, foraging for crabs and possibly fish along receding tide-lines. Nocturnal.

Genus *PSEUDOFERANIA* Monotypic genus.

Macleay's Water Snake

Pseudoferania polylepis

TL 870mm

21–23 midbody scale rows; a loreal scale; divided anal and subcaudal scales; extremely glossy, iridescent sheen. Shades of brown to dark grey, usually with dark ventrolateral stripe, and often vertebral stripe. Ventral surfaces cream to yellow with dark midventral stripe under tail and sometimes on body.

Pseudoferania polylepis. Claudie River.

Tropical freshwater creeks, swamps, rivers of WT, CYP, EIU, GUP, NWH. Also nthn NT, NG. Mainly nocturnal, but sometimes seen foraging by day. Shelters under mats of submerged leaf litter, vegtn, and in cavities along creek banks. Feeds mainly on fish and frogs; shrimps also recorded.

VENOMOUS LAND SNAKES

A Pale-headed Snake (*Hoplocephalus bitorquatus*) lunges and gapes its mouth open when provoked. This arboreal snake feeds largely on frogs. Moranbah area.

Family **Elapidae**; Subfamily **Hydrophiinae**

The most diverse group of Australian snakes are the venomous terrestrial elapids. They occur worldwide, but only in Australia do they dominate the snake fauna in terms of species richness, variation in structure and lifestyle. Terrestrial elapids share a pair of short fixed fangs at the front of the mouth, large, symmetrically arranged head shields, broad expanded ventral scales and cylindrical pointed tails, and have no loreal scales.

They range from species 3m long with venom of unparalleled toxicity to minute, 30cm snakes with mild venom just sufficient to dispatch small skinks. Some are sleek and extraordinarily swift with large eyes and acute vision, able to pursue prey in a burst of speed across open terrain. Others are sluggish and lie immobile amongst leaf litter for days at a time, ready to strike in ambush. Many are exclusively nocturnal, some diurnal, and others both depending on the temperature. The group contains both egglayers and livebearers.

These snakes are now classified in the subfamily Hydrophiinae, reflecting a closer relationship with sea snakes than with terrestrial elapids of other continents.

Qld is rich in terrestrial elapids, with Australia's highest diversity in tropical to subtropical eastern areas.

KEY TO GENERA

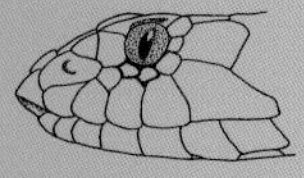

1 Row of scales between eye and upper labials **(a)** .... ***Acanthophis***
No row of scales between eye and upper labials 2

2(1) Anal scale single 3
Anal scale divided 9

3(2) Subcaudal scales divided .... ***Oxyuranus***
Subcaudal scales single 4

4(3) Dorsal scales smooth 5
Dorsal scales keeled .... ***Tropidechis***

5(4) Lips barred 6
No bars on lips 7

6(5) Midbody scales 19–21 .... ***Hoplocephalus***
Midbody scales 17 .... ***Denisonia***

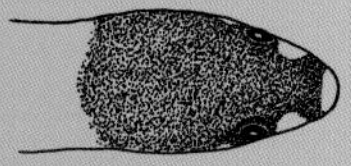

7(5) Lower secondary temporal scale much longer than high **(b)** .... ***Notechis***
Lower secondary temporal scale about as high as long **(c)** .. 8

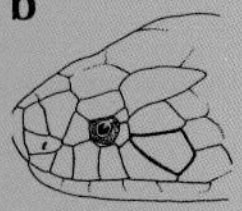

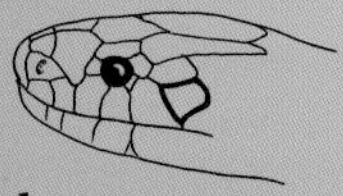

8(7) No dark hood on head, or if present, then continuous with a prominent dark vertebral stripe; no dark spots on head .... ***Cryptophis***
Dark hood **(d)** or dark spots on head; no vertebral stripe (though darker vertebral suffusion may be present) ...... ***Suta***

9(2) All subcaudal scales single .... ***Hemiaspis****
Usually at least some subcaudal scales divided 10

10(9) All subcaudal scales divided (at most a few anterior ones single) 11
At least anterior 20 per cent of subcaudal scales single .... ***Pseudechis****

11(10) Subcaudal scales fewer than 35 pairs 12
Subcaudal scales 35 or more pairs 15

12(11) Black and white rings around body .... ***Vermicella***
Bands present or absent but no rings encircling body 13

13(12) Ventral surfaces pale and patternless 14
Ventral surfaces dark, or marked with narrow dark bands or longitudinal rows of dark blotches or stripe .... ***Cacophis*** (part)

14(13) Rostral scale upturned with an acute cutting edge, or if only weakly protrusive, then body marked with numerous narrow dark bands .... ***Brachyurophis***
Rostral scale weakly protrusive and not upturned; body usually marked with a netted or freckled pattern but no bands .... ***Antaioserpens***

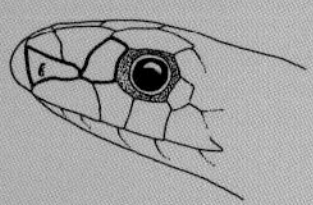

15(11) Nasal and preocular scales separated by contact between prefrontal and upper labial scales **(e)** 16
Nasal and preocular scales in contact **(f)** 18

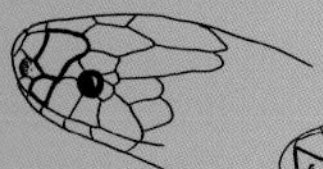

16(15) Belly flecked, typically with orange, but sometimes grey .... ***Pseudonaja*** (part)
Belly pearly white without pattern 17

17(16) Nasal scale not divided .... ***Furina***
Nasal scale divided .... ***Glyphodon***

18(15) Midbody scales in 15 rows.. 19
Midbody scales in 17 or more rows.. ***Pseudonaja*** (part)

19(18) Eye small; pale band or streak on neck; no narrow line across front of snout........ ***Cacophis*** (part)
Eye very large; neck markings present or absent, but if present then a narrow line runs across front of snout between nostrils.. ***Demansia***

*Occasional individuals of *Pseudechis* with all single subcaudal scales would erroneously key out to *Hemiaspis.*

Death Adders
Genus *ACANTHOPHIS*

About 8 spp. in Aust., NG and estn Indonesia; about 6 in Qld. Extremely robust body; abruptly slender tail with segmented tip and a curved spur; a row of scales between eye and upper lips; 19–23 midbody scale rows; single anal scale; single anterior subcaudals and divided posterior subcaudals.

Widespread where shrub and leaf litter layers remain intact in all mainland states with possible exception of Vic. Extremely secretive and sedentary, lying concealed beneath leaf litter with only snout and tail-tip protruding. The segmented tip is wriggled to lure vertebrate prey within striking range. Ambush from a place of concealment, along with the lure and the large, mobile fangs, mirrors many vipers on other continents. Nocturnal to diurnal livebearers.

Though distinctive as a group, some pops with available species names cannot be identified with any certainty. *A. antarcticus* is assumed to cover much of Qld, while death adders in rocky areas of NWH and mesas of MGD are assumed to be *rugosus,* a sp. shared with NG. Those in open wstn downs areas (Barkly Tablelands of NT and adj. Qld) are *hawkei.* These pops are included here under their species names, with the caveat that the descriptions and maps should be treated with a degree of caution. It is noteworthy that this group of viper-like elapids, so distinctive at the generic level, continues to pose problems in the delineation of species. All are DANGEROUS.

Common Death Adder
Acanthophis antarcticus
TL 0.7–1.0m+

Smooth to very weak keels on dorsal scales; smooth to moderately rugose head shields; 21 (occasionally 23) midbody scale rows; moderate to large size, rarely attaining one metre. Grey or reddish brown, generally biased according to substrate colour though both forms can occur together, marked with irregular bands.

Acanthophis antarcticus. Reddish form. D'Aguilar NP.

Acanthophis antarcticus. Grey form. Figtree Stn.

G. Stephenson

Acanthophis antarcticus. Windorah floodplain.

Lips usually prominently barred. Tail-tip cream or black.

Throughout Qld with possible exception of CYP. Also NSW, SA and sthn WA. Generally larger and more robust than praelongus, with flatter peaks over eyes. Formerly abundant in many areas, such as parts of BB, but numbers have declined dramatically. DANGEROUSLY VENOMOUS.

Plains Death Adder; Barkly Tableland Death Adder

Acanthophis hawkei

TL 0.7–1.2m

Smooth moderately keeled dorsal scales; smooth to moderately rugose head shields; 21–23 midbody scale rows; very large size, occasionally exceeding one metre. Pale brown to grey with numerous irregular bands, and dark and pale bars on lips. Very poorly diagnosed, and difficult

J. Farquar

Acanthophis hawkei. Fogg Dam, NT.

to identify solely on morphological characters. Regarded as largest death adder in n. Aust. though reputed maximum sizes are anecdotal.

Far wstn MGD. Also adj. NT to near Darwin. Associated with grasslands on cracking clay. DANGEROUSLY VENOMOUS.

Smooth-scaled Death Adder

Acanthophis laevis

TL 0.5–1.0m

Smooth dorsal scales; moderate to strongly raised peaks above eyes. Yellowish brown to greyish brown with weak to prominent bands across body and prominent, smoothly sharp-edged black blotches on upper lips particularly at rear on 6th and 7th upper labial scales. Ventrolateral row of prominent sharp-edged black blotches often present along body.

Currently known in Qld only from Dauan Is. in nthn Torres Strait. Also NG incl. estn Indonesia. Sole record based on

A. Davies

Acanthophis laevis, Dauan Is.

an image which appears to match this sp., taken in rainforest and palms amongst granite boulders. DANGEROUSLY VENOMOUS.

Northern Death Adder

Acanthophis praelongus

TL 600mm

Moderate keels on anterior back; moderate to strongly rugose head shields; raised supraocular scales forming peaks above each eye; 21–23 midbody scale rows. Reddish brown and grey colour forms occur, each irregularly banded, with lips usually barred, and cream or black tail-tips.

Grasslands, woodlands, rocky ranges of WT, CYP, through Torres Strait to NG. Tends to be smaller and more slender than *A. antarcticus*, with higher peaks over eyes and stronger keels on body scales. DANGEROUSLY VENOMOUS.

Acanthophis praelongus. Badu Is.

Desert Death Adder

Acanthophis pyrrhus

TL 708mm

Strong keels on back and sides; strongly rugose head shields; usually 21 midbody scale rows; relatively slender build. Reddish brown to brick red with numerous cream to yellow bands, mottled lips, usually dark tail-tip.

Acanthophis pyrrhus. Port Hedland area, WA.

Spinifex deserts of CHC. Also NT, SA, WA. DANGEROUSLY VENOMOUS.

Papuan Death Adder

Acanthophis rugosus

TL 0.71m

Keeled anterior dorsal scales; moderate to strongly rugose head shields; 21–23 midbody scale rows; supraocular scales raised to form prominent peaks. Grey to reddish brown with numerous darker bands, dark or pale tail-tip and heavily mottled or barred upper lips. Ventral surfaces marked with distinct dark blotching.

Poorly known, but regarded as primarily associated with extensive rocky areas of NWH to scattered mesas of CHC. Also nthn NT and NG. DANGEROUSLY VENOMOUS.

A. Emmott

Acanthophis rugosus. Dajarra.

Genus *ANTAIOSERPENS*

2 Aust. spp.; both endemic to Qld. Weakly shovel-shaped snouts; prefrontal scale contacts upper labial scales separating nasal scale from preocular scale; smooth, glossy scales in 15 midbody rows; divided anal and subcaudal scales; moderately robust build; short-tails.

Tropical open forests and woodlands of ne., and mulga woodlands of interior. Secretive nocturnal burrowers. Probably feed largely or exclusively on lizards, particularly skinks and possibly lizard eggs. Egg-layers. Regarded as harmless.

Antaioserpens albiceps. Hervey Range.

Antaioserpens warro. Charleville.

North-eastern Plain-nosed Burrowing Snake

Antaioserpens albiceps

TL 420mm

Brown to rich orange, with reticulum formed by darker hind edges to body scales, and sometimes a dark dorsal stripe on tail. Top of head dark grey with pale mottling, separated from broad black blotch on neck by broad to narrow cream to white band. Juv. sometimes more sharply patterned, with nearly completely white head and dark mottling on snout.

Endemic. Open tropical forests and woodlands of CYP, WT, EIU and BB s. nearly to Clermont.

Warrego Burrowing Snake

Antaioserpens warro

TL 438mm

Pale reddish brown to olive speckled with irregular dark-edges to body scales and sometimes a dark dorsal stripe on tail. Blackish hood on top of head extends back and becomes diffuse at about rear edge of parietal scales, separated from broad black band on neck, 6–8 scales long by 3–4 rows of much paler scales.

Endemic. Poplar box/Callitris pine and mulga woodlands in ML and adj. wstn edge of BB. Recorded from Mitchell, Morven area, Charleville, and an original faded holotype from Warro Stn, Port Curtis.

Shovel-nosed Snakes
Genus *BRACHYUROPHIS*

7 Aust. spp.; 4 occur in Qld. Shovel-shaped snout with protruding rostral scale; rostral scale upturned with acute transverse cutting edge and nasal scale contacting preocular scale, or rostral scale not upturned, with weak transverse cutting edge and nasal separated from preocular; smooth, glossy scales in 15–17 midbody rows; divided anal and subcaudal scales; broad dark bands across head and neck; usually narrow prominent bands across body; relatively short tail.

Dry to arid interior and n., penetrating SEQ via well-drained soils. Also dry areas of all mainland States. These nocturnal burrowing snakes employ their shovel-shaped snouts to penetrate soil under embedded stumps and rocks, in cavities and under thick leaf litter. Some specialise on reptile eggs, with dissected specimens containing only collapsed eggs with their soft shells slit. Others eat both eggs and lizards. Egglayers. Regarded as harmless.

KEY TO *BRACHYUROPHIS*

1	Snout upturned, with acute transverse cutting edge; nasal scale contacts preocular scale, separating prefrontal from upper labials **(a)**	2
	Snout not upturned, with very weak transverse cutting edge; nasal separated from preocular by contact of prefrontal with upper labials **(b)**	***fasciolatus***
2(1)	Bands absent on body	***incinctus***
	Bands present on body	3
3(2)	Dark bands composed of dark-edged paler scales	***australis***
	Dark bands composed of wholly dark scales	***campbelli***

Australian Coral Snake
Brachyurophis australis
TL 340mm

Upturned rostral scale with acute cutting edge; nasal scale contacts preocular scale; 17 midbody scale rows. Shades of pink to reddish brown with narrow, ragged-edged bands comprising dark-edged paler scales, and broad dark bands across head and neck.

Dry woodlands, shrublands, outcrops, mainly on sandy soils, in SEQ, BB, ML,

Brachyurophis australis. Moranbah area.

Brachyurophis australis. Capella area.

estn CHC, MGD, DEU, EIU. Also dry areas of NSW, Vic, se. SA. Eats reptile eggs, skinks.

Brachyurophis campbelli. Weipa. *L. Dibben*

Brachyurophis campbelli. Mt Isa.

Brachyurophis campbelli. Almaden. *M. Anthony*

Campbell's Shovel-nosed Snake.

Brachyurophis campbelli

TL 353mm

Upturned rostral scale with acute cutting edge; nasal scale contacts preocular; 15–17 midbody scale rows. Orange to reddish brown with broad dark blotch on top of head (white on specimens from Laura), broad dark band across neck and numerous dark bands across body. The pale interspaces range from much less than one third, to slightly broader than the dark bands.

Endemic. Dry woodlands of western CYP, EIU, NWH, DEU, MGD and GUP. Taxonomic status remains unclear. Specimens from CYP (Laura, Weipa, Wenlock River and Archer River) and Barcaldine have 15 midbody scale rows and those from EIU (Almaden and Mungana) and NWH (Mt Isa) have 17. The name *woodjonesii* (from Lower Archer River; 15 midbody scale rows) is available. Probably a reptile egg specialist.

Narrow-banded Shovel-nosed Snake
Brachyurophis fasciolatus
TL 390mm

Rostral scale not upturned, with relatively weak cutting edge; prefrontal scale broadly contacts upper labials, separating nasal from preocular; 17 midbody scale rows. Cream to very pale grey or pink with prominent dark blotch on head, dark band 3–6 scales wide across neck, numerous small reddish spots alternating with narrow, ragged-edged black bands across body and tail.

CHC. Also dry to arid areas of all mainland States except Vic. Eats reptile eggs, skinks. Represented in Qld by ssp. *B. f. fasciata*. One other ssp. in WA.

Unbanded Shovel-nosed Snake
Brachyurophis incinctus
TL 360mm

Upturned rostral scale with acute cutting edge; nasal scale contacts preocular scale; 17 midbody scale rows; unbanded body. Pink to brown with black blotch on head forward to about level of eyes, broad black band across neck, and reticulum on body formed by dark-edged scales.

Arid woodlands, shrublands, grasslands, mainly on heavy clays, loams, stony soils, in ML, CHC, MGD, NWH. Also NT. Probably reptile egg specialist.

R. Valentic

Brachyurophis fasciolatus. Wombah Gate area.

Brachyurophis incinctus. Mt Isa.

Crowned Snakes
Genus *CACOPHIS*

4 Aust. spp.; all occur in Qld. Pale band or collar on nape; nasal and preocular scales in contact, separating prefrontal from upper labials; smooth glossy scales in 15 midbody rows; divided anal and subcaudal scales.

Estn Qld. Also estn NSW. Variety of habitats, from rainforests to urban parks and gardens, usually in moist shelter-sites under rocks, logs, leaf litter, compost heaps. Nocturnal skink specialists, probably capturing most prey at night while it sleeps. Though disinclined to bite, all have an impressive threat response, rearing head and forebody high off the ground and mock-striking with the mouth closed. Egglayers. Regarded as harmless though bites from largest sp., *C. squamulosus*, could cause discomfort.

KEY TO *CACOPHIS*

1	Complete band across back of neck	2
	Pale streak along each side of neck, failing to form a complete band	***squamulosus***
2(1)	Broad white band, about 4 scales wide, across back of neck	***harriettae***
	Narrow yellow band, no more than 2 scales wide, across back of neck	3
3(2)	Belly pale yellow with narrow dark edges to ventral scales; SEQ to CQC	***krefftii***
	Belly usually dark grey; WT	***churchilli***

Cacophis churchilli. Lake Eacham.

Cacophis harriettae. Warrill View.

Northern Dwarf Crowned Snake

Cacophis churchilli

TL 450mm

Narrow pale yellow band, 1⁄2–2 scales wide, across nape. Glossy dark grey above and usually below. Ventral surfaces occasionally pale yellow with narrow dark bands.

Endemic. Rainforests, eucalypt forests and rock outcrops in WT, s. to Bluewater Ra. and Mt Spec.

White-crowned Snake

Cacophis harriettae

TL 500mm

Broad white band, 4 or more scales wide, across nape; dark ventral surfaces. The white band extends around sides of head to snout, enclosing shiny black patch on top of head. Body glossy dark grey to almost black, above and below.

Very common in moist sheltered sites; particularly successful in urban areas, in SEQ, BB, CQC. Also ne. NSW. Thanks to compost heaps and rockeries, coupled with year-round watering regimes in lush gardens, may now be more successful in some inner Brisbane suburbs than in peripheral bushland.

Cacophis krefftii. Mt Nebo.

Southern Dwarf Crowned Snake

Cacophis krefftii

TL 345mm

Narrow pale yellow band, 1⁄2–2 scales wide, across nape; pale ventral surfaces. Dark grey to almost black above, pale yellow below with narrow dark bands formed by dark edges to ventral scales.

Rainforests, sheltered moist sites in dry forests, urban areas of SEQ, coastal BB, CQC. Also estn NSW. Occurrence patchy. Frequently encountered in some Brisbane suburbs (e.g. Bardon, Indooroopilly), yet rarely in others.

Cacophis squamulosus. Mt Glorious.

Golden-crowned Snake

Cacophis squamulosus

TL 750mm

Pale streak along each side of neck failing to form complete band; orange ventral surfaces. Dark brown to dark grey with pale yellow to pale grey or brown streak extending back each side of neck, and forward to enclose snout. Ventral surfaces with prominent row of brown blotches along midline.

Variety of well-watered habitats along coast and ranges of SEQ, NET, coastal BB, CQC, and isolated inland records from Carnarvon Ra. in BB. Also estn NSW. Largest *Cacophis*. Bite could cause discomfort but, like its relatives, impressive threat display is largely bluff and mouth is rarely opened.

Genus *CRYPTOPHIS*

5 Aust. spp.; 4 occur in Qld. Relatively depressed heads; squarish snouts; very small to moderate-sized eyes with dark irises; smooth glossy scales in 15 midbody rows; single anal and subcaudal scales; pattern absent or reduced to dark head and vertebral stripe.

Estn Qld, from SEQ to Torres Strait. Also nthn and estn Aust. to sthn NG. Nocturnal, sheltering by day under rocks, logs, surface debris and in soil cracks. Most are probably lizard specialists, though *C. nigrescens* also takes snakes and frogs, and is sometimes cannibalistic. Livebearers. Most are not considered dangerous but *C. nigrescens* has been implicated in a fatality and many other serious bites. All should be regarded as POTENTIALLY DANGEROUS.

KEY TO *CRYPTOPHIS*

a **b**

1 Prefrontal scales in broad contact with upper labial scales **(a)** .... ***boschmai***
Nasal and preocular scales in contact, separating prefrontals from upper labial scales **(b)** 2

2(1) Uniformly dark above .... ***nigrescens***
Dorsal colour incl. brown or pink 3

3(2) Uniformly pink .... ***incredibilis***
Dark head and dark vertebral stripe .... ***nigrostriatus***

Cryptophis boschmai. Oakey area.

Carpentaria Snake
Cryptophis boschmai
TL 560mm

Prefrontal scales in broad contact with upper labial scales; little or no pattern. Shades of brown with dark scale-bases, sometimes dark flush along midline.

Endemic. Dry woodlands, mainly w. of coastal ranges, in SEQ, BB, CQC, DEU, EIU, WT, nthn CYP.

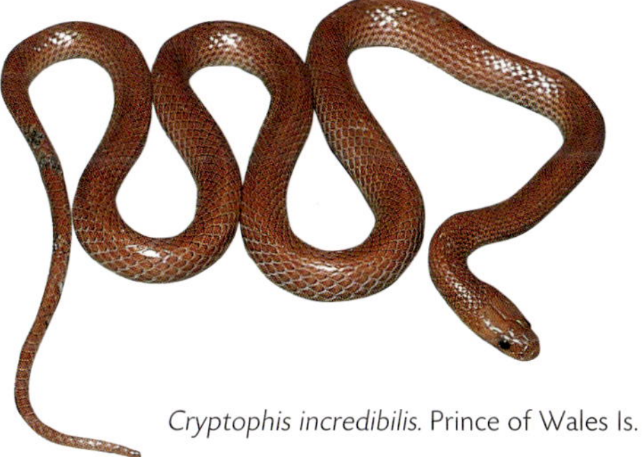

Cryptophis incredibilis. Prince of Wales Is.

Pink Snake
Cryptophis incredibilis
TL 400mm

Nasal and preocular scales in contact, separating prefrontals from upper labials; relatively large eyes; slender body with long tail. Uniformly pink above, pearly white below.

Endemic. One of most restricted distributions of any Qld snake. Known only from Prince of Wales Is. in sthn Torres Strait; inhabits woodlands on sandy soils.

Cryptophis nigrescens. Mt Mee.

Small-eyed Snake
Cryptophis nigrescens
TL 500mm (1m max.)

Nasal and preocular scales in contact, separating prefrontals from upper labials; small eyes; no pattern. Shiny black to bluish grey above and pink, occasionally cream, below. Pale ventral colours entirely restricted to belly (unlike superficially similar Red-bellied Black Snake, *Pseudechis porphyriacus*, which has intense pink to red on the lower flanks).

Mainly associated with forests, from rainforests to moist retreats within dry sclerophyll forests, in NET, SEQ, BB, CQC, WT. Also estn NSW, Vic. One recorded fatality and numerous other serious bites. DANGEROUSLY VENOMOUS.

Black-striped Snake
Cryptophis nigrostriatus
TL 500mm

Nasal and preocular scales in contact, separating prefrontals from upper labials; relatively large eyes; slender body with long tail; black vertebral stripe. Shades of brown with black to dark brown head continuous with broad dark stripe to tail-tip, pearly white below.

Dry forests, woodlands and rock outcrops of coastal BB, CQC, EIU, WT, CYP. Also sthn NG.

Cryptophis nigrostriatus. Glenden area.

Whipsnakes

Genus *DEMANSIA*

13 Aust. sp.; 8 occur in Qld. Large prominent eyes with pale irises and round pupils; usually dark teardrop or comma shape below eye; very slender bodies with long, finely tapering tails; smooth, matt-textured scales in 15 midbody rows; divided anal and subcaudal scales.

Throughout Qld, mainly in dry open areas. These extremely swift, alert, diurnal snakes are essentially the cheetahs of the snake world, employing acute vision and a burst of speed to capture swift, active lizards. Egglayers. Bites are not normally serious but 2 largest spp. should be considered POTENTIALLY DANGEROUS.

KEY TO *DEMANSIA*

1	Dark spots on anterior ventrals, diverging as they extend back	***rimicola***
	Anterior ventrals patternless or with dark posterior edge	2
2(1)	One or more dark nuchal bands (fading with age)	3
	No dark nuchal bands on juveniles or adults	5
3(2)	Dark nuchal band(s) with distinct pale edge; when faded, some indication of pale edge persists along on sides of neck	4
	Dark nuchal band with no pale edges; NWH only	***quaesitor*** (part)
4(3)	Two dark nuchal bands	***flagellatio***
	One dark nuchal band	***torquata***
5(2)	Ventral scales mostly dark grey; no dark line across snout (weak if present, and lacking pale edges)	6
	Ventral scales pale yellow to cream; pale-edged dark line across snout	7
6(5)	Ventral scales fewer than 197	***vestigiata***
	Ventral scales more than 198	***papuensis***
7(5)	Dark comma under eye short, ending on 5th upper labial scale	8
	Dark comma under eye long, extending across 5th upper labial scale and well onto 6th	***quaesitor*** (part)
8(7)	Head and tail flushed with copper; far sw. Qld	***reticulata***
	Head and tail not flushed with copper; widespread over estn half of state	***psammophis***

G. Stephenson

Demansia flagellatio. Lawn Hill NP.

Ornate Whipsnake

Demansia flagellatio

TL 0.72m

Dark head and two dark bands across neck; dark line across front of snout; extremely slender build; very long thin tail. Brown to reddish brown, flushed with yellow on tail, with pale-edged dark line across snout not reaching eye, broad black comma-shape under eye sweeping back to corner of mouth, and dark cap on head, separated from two black bands across neck by orange to yellow interspaces. Tail up to 40 per cent of SVL.

Endemic. NWH, on stony soils with extensive spinifex.

E. Vanderduys

Demansia papuensis. Georgetown area.

Greater Black Whipsnake

Demansia papuensis

TL 1.65m

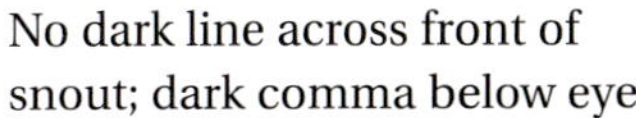

No dark line across front of snout; dark comma below eye very short to absent; 198–228 ventral scales; very large size. Black or dark grey to tan, often flushed with red towards tail, sometimes with narrow broken pale margin around eye and often irregular scattered dark spots on top of head.

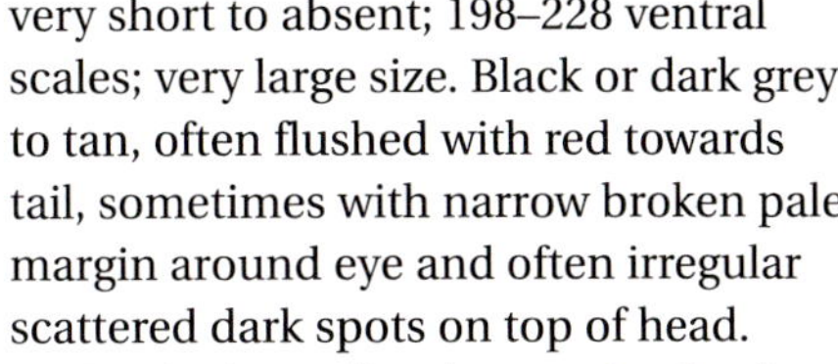

Tropical woodlands, grasslands of NWH, GUP, CYP, EIU, WT, DEU, nthn BB. Also nthn NT and WA, but not NG despite name. POTENTIALLY DANGEROUS.

Yellow-faced Whipsnake

Demansia psammophis

TL 1m

Thin, pale-edged dark line across front of snout between nostrils; pale-edged dark comma below eye sweeps back toward corner of mouth but does not reach last upper labial scale. Pale grey, bluish grey to olive, often with 2 reddish brown paravertebral stripes from behind neck to anterior third of body.

Demansia psammophis. Kurwongbah.

Demansia psammophis. Kurwongbah.

Occupies variety of habitats, from edges of rainforests and vine thickets to open forests, woodlands, disturbed urban areas through SEQ, NET, BB, CQC, EIU and WT. Common in suburban Brisbane gardens. During late spring females sometimes aggregate to lay their eggs communally.

Sombre Whipsnake

Demansia quaesitor

TL 0.74m

Usually a thin dark line, with or without pale-edge, across front of snout between nostrils, narrowly reaching eye; head flushed with copper; usually no dark band across nape (traces laterally on juv.). Greyish brown with copper head and dark comma from below eye to corner of mouth. Specimen with dark nape band, lacking pale edges, recorded from Bowthorne Stn near Doomadgee. Ventral surfaces patternless – at most a few dark flecks on throat.

NWH and MGD. Also nthn NT and WA. Variety of open habitats and creek edges. Status of Qld pops lacking dark band across nape is uncertain as true *D. quaesitor* (NT and WA) has nuchal band.

Demansia quaesitor. Mt Isa.

Demansia quaesitor. Mt Isa.

G. Swan

Demansia reticulata. Nappa Merrie Stn.

Reticulated Whipsnake

Demansia reticulata

TL 1m

Thin, pale-edged dark line extends across front of snout between nostrils; pale-edged dark comma below eye sweeps back toward corner of mouth but does not reach last upper labial scale; dark edged scales on body. Greenish grey with fine reticulum over body and copper flush on head and tail. Represented in Qld by ssp. *D. r. cupreiceps*.

Shrublands and red dunes with spinifex in arid CHC. Also arid zones of SA, sthn NT and WA.

Blacksoil Whipsnake

Demansia rimicola

TL 0.98m

Weak dark line across front of snout, often absent centrally and not reaching eye; pale-edged dark band across nape; dark anterior ventral

Demansia rimicola. Julia Creek.

Demansia torquata. Gladstone.

spots. Brown to yellowish brown, with pigment tending to concentrate to form a series of reddish lateral stripes. Dark band across neck often fading on adults, with pale edges persisting. Dark comma curves back from below eye to corner of mouth, edged by pale bars, with posterior bar constricted to form isolated blotch. Ventral surfaces often yellow to bright red, with dark anterior ventral spots diverging as they extend back.

MGD, CHC, NWH and ML, extending e. to near Roma in BB. Also nthn NT and ne. WA. Favours open grasslands, particularly Mitchell grasses, on cracking clay soils.

Collared Whipsnake

Demansia torquata

TL 0.83m

Pale-edged dark band across nape; prominent pale-edged dark line across front of snout reaching eyes. Olive brown to greyish brown with top of head darker to black. Broad black bar edged with cream to yellow across neck, fading to virtual absence on large adults. Prominent pale-edged dark comma extends back from below eye to behind corner of mouth. Ventral surfaces bluish grey, flushed with yellow under throat and tail.

Endemic. Variety of open habitats in CQC, WT, CYP, and extending inland to Injune area in BB.

Lesser Black Whipsnake

Demansia vestigiata

TL 1.2m

No dark line across front of snout; dark comma below eye very short to absent; 165–197 ventral scales; very large size. Dark brown, reddish brown to dark grey, often flushed with reddish brown towards tail, with narrow broken pale margin around eye and usually small dark blotches on top and sides of head. Body marked with netted or spotted pattern formed by dark or pale markings on individual scales.

Tropical woodlands, grasslands of CYP, GUP, EIU, WT, BB, SEQ, extending s. to coastal heaths of Sunshine Coast, woodlands of Ipswich area. Also nthn NT, WA, sthn NG. POTENTIALLY DANGEROUS.

Demansia vestigiata. Moranbah area.

Genus *DENISONIA*

2 Aust. spp.; both occur in Qld. Very robust build; pale irises and vertically elliptic pupils; smooth scales in 17 midbody rows; single anal and subcaudal scales; barred lips.

Low-lying areas, mainly on heavy, cracking soils subject to flooding in mid-east and interior. Also NSW. Secretive nocturnal snakes that shelter by day in soil cracks and in well-insulated sites under debris. They feed almost entirely on frogs, rarely lizards. Livebearers. Normally very sedentary; when threatened they flatten the body, often with the head hidden beneath coils, but will thrash and strike readily if provoked. Severe local symptoms have been recorded from one sp., so both should be treated with caution.

De Vis' Banded Snake; 'Mud Adder'

Denisonia devisi

TL 568mm

Dorsal pattern present. Shades of brown to olive with large dark blotch on top of head and numerous ragged, dark brown bands, sometimes broken along midline anteriorly.

River edges and low-lying, seasonally inundated sites in BB, ML, MGD, estn CHC. Also interior of NSW.

Denisonia devisi. The Gums.

Ornamental Snake

Denisonia maculata

TL 424mm

Little or no dorsal pattern; shades of grey with darker patch on top of head and black flecks or spots along outer edges of throat and ventral scales.

Endemic. Associated with low-lying, seasonally flooded areas of BB from Charters Towers area s. to Rockhampton and inland to about Blackwater. High densities, often very localised, occur at suitable sites. One recorded bite produced severe local swelling. POTENTIALLY DANGEROUS.

Denisonia maculata. Moranbah area.

Genus *FURINA*

3 Aust. spp.; all occur in Qld. Prefrontal scales broadly contacting upper labials, widely separating nasals from preoculars; undivided nasal scales; weakly flattened heads; small dark eyes; smooth, glossy scales in 15 midbody rows (within Qld but 17 in some NT and WA pops); divided anal and subcaudal scales. Pattern, when present, comprises red to orange nuchal blotch or band.

Widespread, favouring dry areas and tending to shun rainforests, wetlands and other moist habitats. Nocturnal snakes, sheltering in soil cavities and under debris, emerging at night to capture sleeping diurnal lizards, mainly skinks. Identification can be problematic as the main diagnostic feature is not always consistent; whether the red nuchal marking is lunate and largely or fully enclosed within black head and neck, or forms a transverse bar bisecting the black pigment. Not regarded as harmful but care should be taken with large individuals.

KEY TO *FURINA*

1 Red mark on neck usually lunate or oval, tapering towards outer edges and largely enclosed within or just encroaching beyond dark pigment........ ***diadema***
Red, orange to yellow mark on neck if present, forms a band broadly bisecting dark pigment........ 2

2 Preocular scales usually contact frontal scale; pale nuchal band usually discernible, sometimes excluded from parietal scales or extending onto rear portions........ ***ornata***
Preocular scales separated from frontal or just in point contact; pale nuchal band often absent on large adults; when present often extends well onto parietal scales........ ***barnardi***

Yellow-naped Snake
Furina barnardi
TL 500mm

Diffuse pale band across nape; preocular scales separated from frontal or in very narrow point contact. Shiny orange brown to dark brown or almost black, scales sometimes with pale edges or bases creating an obscure reticulum or lines, with a broad pale brown, orange to yellow band across nape and base of head, often extending well onto parietal scales. This darkens with age and may be completely absent on large individuals.

Endemic. Status uncertain. Poorly known, from dry woodlands, rock outcrops in widely separated localities in WT, EIU, BB, nthn CHC, inland to about Opalton.

Furina barnardi. Large adult lacking nuchal band. Eversfield Stn.

Furina barnardi. Cobbold Gorge.

Furina diadema. Injune.

Furina diadema. Bendidee SF.

Red-naped Snake

Furina diadema

TL 400mm

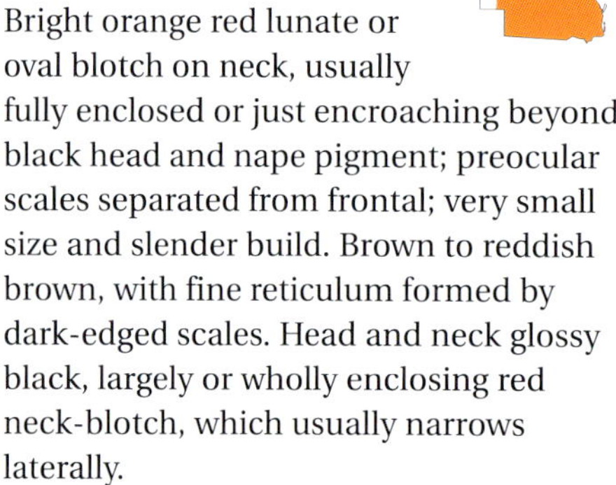

Bright orange red lunate or oval blotch on neck, usually fully enclosed or just encroaching beyond black head and nape pigment; preocular scales separated from frontal; very small size and slender build. Brown to reddish brown, with fine reticulum formed by dark-edged scales. Head and neck glossy black, largely or wholly enclosing red neck-blotch, which usually narrows laterally.

Wide range of dry habitats across all bioregions, with probable exception of CYP, possibly GUP. Also NSW, Vic, SA. One of most commonly encountered small snakes in suburban gardens of Ipswich. Inoffensive and disinclined to bite.

Orange-naped Snake

Furina ornata

TL 650mm

Broad orange-red band across neck; moderately small size and slender build; preocular scales contact frontal. Shades of brown to reddish brown, with a fine reticulum formed by dark-edged scales. Head and neck glossy dark brown to black, broken by a broad, pale neck-band, its edges not becoming constricted laterally.

Woodlands, outcrops from nthn BB and DEU to EIU, WT, GUP, NWH, CYP, n. to sthn Torres Strait. Also NT, nthn SA, WA.

Furina ornata. El Questro, WA.

Furina ornata. El Questro, WA.

Genus *GLYPHODON*

2 Aust. spp.; both occur in Qld. Moderately robust build; weakly depressed heads with small black eyes; prefrontal scales broadly contact upper labials separating nasals from preocular scales; nasal scales divided; smooth, very glossy scales in 17–21 midbody rows; anal and subcaudal scales divided.

Woodlands, vine thickets and monsoon forests of subtropical estn interior and tropical n. Secretive, nocturnal snakes which shelter in soil cracks and other cavities and under surface debris. They probably feed exclusively on skinks. Bites could potentially cause severe symptoms.

Glyphodon dunmalli. Yuleba area.

Glyphodon tristis. Iron Range NP. *A. Zimny*

Dunmall's Snake

Glyphodon dunmalli

TL 700mm

Little or no pattern; 21 midbody scale rows; robust build. Shades of brown to olive with only a few diffuse pale blotches on upper lips.

BB, in vegtn communities incl. brigalow, belah and cypress pine, usually on heavy soils. Also adj. NSW. Extremely secretive, rarely encountered, possibly genuinely scarce. POTENTIALLY DANGEROUS.

Brown-headed Snake

Glyphodon tristis

TL 1m

Broad, diffuse, pale yellowish brown band across neck and base of head; 17 midbody scale rows; robust build. Dark purplish brown to almost black, with pale edges to scales forming reticulum, boldest on lower flanks.

Woodlands, monsoon forests, vine thickets in far nthn EIU and CYP, n. through Torres Strait to sthn NG. No serious bites recorded but because of size and nervous disposition should be regarded as POTENTIALLY DANGEROUS.

Genus *HEMIASPIS*

2 Aust. spp.; both occur in Qld. Large eyes with round pupils and dark brown iris with pale upper edge; smooth, weakly glossed scales in 17 midbody rows; divided anal scale; single subcaudal scales.

Moist areas along e. coast and seasonally flooded parts of drier estn interior. Also estn NSW. Livebearers. One is a nocturnal frog predator, the other a diurnal opportunist taking frogs and skinks. Not considered dangerous, though care should be taken with large individuals.

Grey Snake

Hemiaspis damelii

TL 600mm

Dark band on nape and base of head; patternless face; pale belly. Grey to olive with dark patch on top of head and first few scale rows on nape of juv., reducing to base of head and neck on adult, ventral surfaces cream.

Favours cracking flood-prone soils in BB, extending e. to Lockyer Valley, SEQ. Also ne. interior of NSW. Shelters in soil cavities and beneath well-insulated debris. Largely nocturnal. Feeds on frogs.

Hemiaspis damelii. Bendidee SF.

Black-bellied Swamp Snake; Marsh Snake

Hemiaspis signata

TL 700mm

Two narrow pale stripes on each side of face; dark belly. Brown, olive to black, with darkest snakes often having pale yellow flush on head. Facial stripes prominent, extending from snout or eye to side of neck, and from snout through upper lip to corner of mouth. Ventral surfaces dark grey to black.

Well-vegetated, moist areas such as rainforests, wet sclerophyll forests, edges of waterways, occurring as 3 disjunct pops in WT, CQC, SEQ. Also estn NSW. Mainly diurnal, though active at night in hot weather. Shelters under thick vegtn and debris. Feeds on skinks, frogs.

Hemiaspis signata. Brisbane.

Broad-headed Snakes
Genus *HOPLOCEPHALUS*

3 Aust. spp.; 2 occur in Qld. Broad flat head, distinct from neck; moderately large eye with pale iris and round pupil; smooth, weakly glossed scales in 19–21 midbody rows; laterally keeled or notched ventral scales; single anal and subcaudal scales; barred lips.

Se. coastal ranges and estn interior. Also estn NSW. Arboreal and rock-inhabiting, but unlike true tree snakes (Colubridae) they mainly occupy the equivalent to terrestrial niches, being found in cavities behind loose bark, in hollows and amongst thick epiphytes rather than amongst foliage or on slender boughs. Livebearing. If provoked they rear the forebody, flatten the head and strike repeatedly. No known fatalities, but severe symptoms recorded so they should be considered POTENTIALLY DANGEROUS.

Pale-headed Snake
Hoplocephalus bitorquatus
TL 800mm

Pale band across neck; patternless body; 19–21 midbody scale rows. Normally grey, but can be darker to almost black, with broad paler grey to white band across neck margined in front by dark blotches and behind by continuous to broken black bar. Top of head pale grey, often with darker blotches. Lips normally with dark and pale bars.

Dry sclerophyll forests and woodlands, particularly ironbark, often near water-courses or flood-prone areas in BB, EIU, DEU, MGD, ML and SEQ. Also ne. NSW. Usually associated with mature trees featuring hollows and/or peeling bark; often encountered at night foraging on rough trunks. Eats mainly frogs, also lizards, small mammals. POTENTIALLY DANGEROUS.

Hoplocephalus bitorquatus. Lake Broadwater.

Stephens' Banded Snake
Hoplocephalus stephensii
TL 1.2m

Body usually banded; face marked with dark bars or blotches; 21 midbody scale rows. Variable colouration, but usually dark grey with numerous ragged-edged brown, orange-brown to cream bands, brightest on anterior body. Some pops (and occasional individuals throughout range) unbanded, but facial pattern always remains.

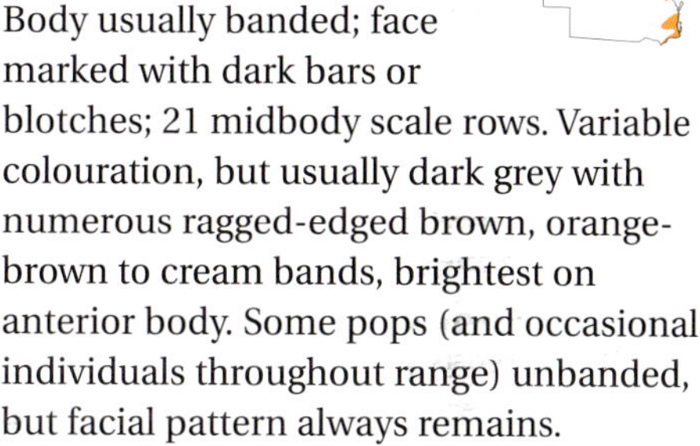

Rainforests, wet sclerophyll forests, rock outcrops in NET, SEQ, n. to Kroombit Tops. Also ne. NSW. Rainforest pops normally inhabit large mature trees with hollows and extensive vine cover and rarely descend to the ground. Rock-inhabiting pops (mainly on granite or sandstone) shelter under slabs and in deep crevices. Eats lizards, frogs, mammals, birds. POTENTIALLY DANGEROUS.

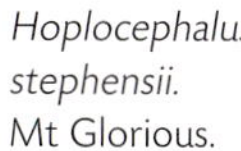

Hoplocephalus stephensii. Mt Glorious.

Genus *NOTECHIS*

Monotypic genus (considered by some to comprise 2 spp.); 1 occurs in Qld.

Tiger Snake

Notechis scutatus

TL 1–2m

Broad, relatively flat head; frontal shield about as wide as long; moderately large eye with dark iris and round, narrowly pale-edged pupil; smooth, weakly glossed scales in 17–21 midbody rows; single anal and subcaudal scales; large size, robust build. Shades of brown or olive to almost black, normally with numerous ragged-edged pale bands, ventral surfaces cream to grey.

SEQ, BB. Widespread but patchy in SEQ. Most pops (McPherson Ra. and Bunya Mtns) occur in upland rainforests. Occupies coastal wallum and heath in lowlands of Sunshine Coast and hinterland, n. to about Maryborough. Isolated pop. occurs in Mt Moffatt sector of Carnarvon NP. Also NSW, all sthn States. Usually associated with creek banks, other moist sites. Diurnal. Livebearer. Dietary opportunist, eating most small vertebrates. Responsible for many human fatalities. DANGEROUSLY VENOMOUS.

Notechis scutatus. Springbrook.

Taipans

Genus *OXYURANUS*

3 Aust. spp.; 2 occur in Qld. Long narrow head; large eyes with round pupils; 21–23 midbody scale rows; single anal scale; divided subcaudal scales; very large size.

Estn, nthn and sw. Qld. Also across nthn Aust., sthn NG, NSW, ne. SA. Taipans are famous for their extremely toxic venom, ranking them amongst the world's most lethal snakes. Both spp. specialise on mammals, with juv. preying on young mice, adults mainly taking rats. Prey is attacked with a rapid (and sometimes multiple) 'strike and release', and followed until it succumbs to the venom and can be eaten at leisure. Egglayers. Diurnal, often foraging during warm to hot mornings. DANGEROUSLY VENOMOUS.

Oxyuranus microlepidotus. Dark winter colouration. Morney Plain.

L. Allen

Oxyuranus scutellatus canni. Saibai Island.

Oxyuranus microlepidotus. Pale summer colouration. Goyder's Lagoon, SA.

Oxyuranus scutellatus scutellatus. North Qld.

Western Taipan; Inland Taipan; Small-scaled Snake; 'Fierce Snake'

Oxyuranus microlepidotus

TL 2m

Dark iris; evenly rounded brow; smooth scales in 23 midbody rows. Colour varies seasonally. In summer snakes are pale brown to yellowish brown with glossy dark heads, and dark edges to body-scales forming herringbone pattern. In winter they become darker, rich brown to almost black, tending to mask the pattern. Ventral surfaces cream to yellow with orange blotches.

Featureless, sparsely vegetated plains of CHC. Also adj. SA and NSW. At night and during hot periods, shelters in deep soil cracks and burrow systems of its major prey, the Long-haired Rat (*Rattus villosissimus*), emerging to bask for brief periods. In rare bountiful seasons, when rat numbers reach plague proportions, offering limitless food resources, taipans are numerous, fat and sleek.

Based on test results from mice, venom of *O. microlepidotus* may be most toxic known for any terrestrial snake. Interpretation of test results may be biased, as the snake is specialised to prey on the very animal group tested. Despite having the common name 'Fierce' Snake it is shy and no more inclined to bite or attack than any other snake. DANGEROUSLY VENOMOUS.

Coastal Taipan

Oxyuranus scutellatus

TL 2m (individuals up to 3m recorded)

Orange-brown iris; long head with angular brow; smooth scales (weakly-keeled on neck) in 21–23 midbody rows. Ssp: *O. s. scutellatus* is pale brown to almost black, with distinctly paler head. This often darkens with age but is always discernible at least on snout. Ventral surfaces cream, with orange spots or blotches. *O. s. canni* normally has broad orange vertebral stripe.

O. s. scutellatus occurs in grassy woodlands and canefields along coast and ranges in CYP, WT, EIU, BB and SEQ. Also ne. NSW and across nthn Aust. *O. s. canni* occurs on nthn Torres Strait is. and sthn NG. Shelters in hollow logs and abandoned burrows; feeds mainly on rats but recorded to take prey as large as bandicoots. Extremely alert, with acute vision and, according to those who work closely with them, apparently high intelligence. Nervous, easily provoked if cornered or harassed, but quick to flee if given the opportunity. DANGEROUSLY VENOMOUS.

Black Snakes

Genus *PSEUDECHIS*

8 spp. in Aust. and NG; 6 occur in Qld. Broad, depressed heads; smooth, weakly to strongly glossed scales in 17–19 midbody rows; usually a divided anal scale; at least anterior 20 per cent of subcaudals single and remainder divided (often all single on one sp); moderate to very large size and robust build.

Throughout Qld, most of Aust. and sthn NG. Mainly nocturnal during hot weather, diurnal during mild weather. Terrestrial, sheltering in abandoned burrows, soil cracks and hollow logs. All are opportunistic predators, taking a broad variety of vertebrate prey. Most are egglayers, but one sp. bears live young, born in membranous sacs. Bites from some spp. have caused fatalities. All should be considered DANGEROUSLY VENOMOUS.

KEY TO *PSEUDECHIS*

1	Midbody scales in 17 rows	2
	Midbody scales in 19 (rarely 21) rows	4
2(1)	Black above; iris black	***porphyriacus***
	Brown above; iris reddish brown	3
3(2)	Head broad; no streaks on neck; TL to 2m; usually most subcaudal scales (posterior 80 per cent) divided	***australis***
	Head relatively narrow; neck usually streaked; TL to 1.2m; subcaudal scales often all single	***pailsei***
4(1)	Ventral scales 200 or fewer	***guttatus***
	Ventral scales 205 or more	5
5(4)	Body blotched or banded; MGD	***colletti***
	Body patternless; nthn Torres Strait	***papuanus***

Pseudechis australis. Glenmorgan.

Pseudechis colletti. Richmond area.

King Brown Snake; Mulga Snake

Pseudechis australis

TL 2–2.5m

Reddish brown iris; 17 midbody scale rows; brown to rich reddish brown, usually with netted pattern formed by pale bases and darker hind-edges to each scale, ventral surfaces cream to white.

Occurs in most dry, open habitats, avoiding moist closed forests and wetlands, in all bioregions except SEQ, NET. Also NG, all mainland States except Vic. Egglayer. One of largest Australian venomous snakes, with highest venom yield. DANGEROUSLY VENOMOUS.

Pseudechis guttatus. Glenmorgan.

Collett's Snake

Pseudechis colletti

TL 1.5m

Dark iris; 19 midbody scale rows. Grey to reddish brown, merging to pink on mid to lower flanks, with large, irregular, pink to red bands or transversely aligned blotches, ventral surfaces cream to pink.

Endemic. Restricted to open grasslands on cracking clay soils in MGD. Shelters in deep soil cracks. Egglayer. DANGEROUSLY VENOMOUS.

Spotted Black Snake; Blue-bellied Black Snake

Pseudechis guttatus

TL 1.5m

Dark iris; 19 midbody scale rows; dark belly. Variable in colour. Typically grey to black with small, scattered pale spots. Some are dark and patternless, some cream with black-tipped scales, others flushed with brown or pink and superficially resembling *P. colletti.* Ventral surfaces grey to bluish grey.

Often associated with river floodplains, temporary wetlands supporting woodlands and dry sclerophyll forests, incl. brigalow communities, in BB, NET, SEQ, e. to about Lockyer Valley. Also ne. NSW. Egglayer. DANGEROUSLY VENOMOUS.

Pseudechis pailsei. John Hills via Winton.

Pseudechis porphyriacus. Bunya Mtns.

Eastern Pygmy Mulga Snake

Pseudechis pailsei

TL 1.2m

Dark reddish brown iris; 17 midbody scale rows; subcaudal scales often all single; usually dark sutures on head and dark streaks on neck; relatively slender build with narrow head. Brown to reddish brown above and cream below, with dark rear edges to body scales creating a very weak reticulum. Longitudinal dark streaks usually present on neck, and usually faint dark edges to dorsal head sutures. Status uncertain and possibly conspecific with *P. weigeli* of nthn NT and WA.

Presumed endemic to NWH and ranges of wstn MGD, occurring in areas of rock and spinifex. Egg-layer. Assumed to be DANGEROUSLY VENOMOUS.

Red-bellied Black Snake

Pseudechis porphyriacus

TL 1.5–2m

Dark iris; 17 midbody scale rows. Uniform glossy black above with brown snout and dull to bright red lower flanks, continuous with red to pink belly.

Swamps, riverbanks, rainforests in SEQ, BB, NET; disjunct pops in CQC, WT. Also NSW, Vic, se. SA. Livebearing. Young born in membranous sacs, the rudiments of shelled eggs, from which they emerge within minutes. DANGEROUSLY VENOMOUS.

Papuan Black Snake

Pseudechis papuanus

TL 2.1m

Pale brown iris; 19 (rarely 21) midbody scale rows. Black to dark brown, sometimes with reddish ventral flush on chin, posterior body and tail. Ventral surfaces dark grey.

Recorded from Saibai Island, a low-lying mosaic of freshwater and saline wetlands just s. of NG mainland in far nthn Torres Strait. Widespread in sthn NG. Egglayer. DANGEROUSLY VENOMOUS.

L. Allen

Pseudechis papuanus. Saibai Island.

Brown Snakes
Genus *PSEUDONAJA*

9 Aust. spp.; 7 occur in Qld. Relatively narrow heads; moderately large eyes with round pupils and usually pale irises (dark in one sp.); smooth, weakly glossed scales in 17–21 midbody rows; divided anal and subcaudal scales (at most a few anterior subcaudals single); orange spots on belly (occasionally grey on dark individuals); black head-blotch and nape-band on juv.

Throughout Qld, mainly in dry, well-drained areas. Also dry parts of all mainland States, sthn NG. Terrestrial, sheltering in abandoned burrows, soil cracks and hollow logs. Opportunistic predators, combining envenomation and constriction to subdue vertebrates. Nocturnal to diurnal according to temperature. Egglayers. Normally shy, but ill-tempered if cornered or molested, rearing the forebody, gaping the mouth and striking. Included are some of Qld's most abundant and lethal snakes. With the possible exception of one small sp. (*P. modesta*), all are DANGEROUSLY VENOMOUS.

KEY TO *PSEUDONAJA*

1 Midbody scales 21 (rarely 19) .... ***guttata***
Midbody scales 17 2

2(1) Ventral scales fewer than 185 .... ***modesta***
Ventral scales more than 185 3

a b

3(2) Mouth-lining flesh-coloured 4
Mouth-lining blackish 5

4(3) Snout rounded from above **(a)** .... ***textilis***
Snout squared or 'chisel-shaped' from above, formed by broad, strap-like rostral scale **(b)** .... ***aspidorhyncha***

5(3) Iris very dark and difficult to determine; cracking clay soils of wstn MGD .... ***ingrami***
Iris includes a clearly visible (sometimes broken) orange circle around pupil; tropical woodlands or sandy deserts 6

6(5) Snout squared or 'chisel-shaped' from above, formed by broad, strap-like rostral scale (b); tropical woodlands .... ***nuchalis***
Snout rounded from above (a); sandy deserts .... ***mengdeni***

Strap-snouted Brown Snake
Pseudonaja aspidorhyncha
TL 1.3 m

Midbody scales in 17 rows; pale mouth-lining; dark iris with pale orange ring around pupil; large strap-like rostral scale creating squared snout in dorsal view. Brown to greyish brown exhibiting numerous, extremely variable pattern combinations. These incl.: scattered darker scales; broad darker band on neck; black V- or W-shape on neck; darker head; numerous darker bands ranging from indistinct, narrow and most prominent laterally and posteriorly to broad, conspicuous and sometimes

Pseudonaja aspidorhyncha. Mitchell area.

Pseudonaja aspidorhyncha. Morven area.

alternating with narrower dark bands within pale interspaces.

CHC, MGD, ML, BB and possibly sthn DEU. Also wstn NSW, wstn Vic and estn SA, but distributional limits ill-defined pending assessment of museum specimens currently assigned to *P. nuchalis*. Dry to arid areas, particularly woodlands and shrublands on heavy soils. DANGEROUSLY VENOMOUS.

Speckled Brown Snake
Pseudonaja guttata
TL 1.4m

Midbody scales in 21 (rarely 19) rows; bluish black mouth-lining; reddish yellow iris, inner margin narrowly edged with contrasting white; no prominent black head and nape markings on juv. Pale greyish brown to orange-yellow, with speckled pattern formed by black lateral edges to most scales. Some snakes have up to 12 evenly spaced broad dark bands, sometimes with 3 narrow transverse lines in between.

Open grasslands, shrublands of MGD, CHC. Also adj. NT, SA. Shelters in deep soil cracks. DANGEROUSLY VENOMOUS.

Pseudonaja guttata. East Lynne Stn, Eromanga area.

Pseudonaja guttata. Morney Plain.

Ingram's Brown Snake
Pseudonaja ingrami
TL 1.8m

Midbody scales in 17 rows; bluish black or bluish grey mouth-lining; dull orange-brown iris, superficially appearing black; 7 lower labial scales. Shades of brown, often with dark-tipped scales, and sometimes contrastingly darker head and forebody, particularly on pale-coloured snakes.

S. Eipper

Pseudonaja ingrami. Boulia.

Pseudonaja mengdeni. Ethabuka.

A. Emmott

Pseudonaja ingrami. Bedourie area.

Pseudonaja mengdeni. Nappa Merrie Stn.

Ventral surfaces pale to bright yellow with paired series of red to orange spots.

Low-lying, seasonally flooded areas in MGD, CHC. Also mid e. NT, far ne. WA. Shelters in deep soil cracks. DANGEROUSLY VENOMOUS.

Western Brown Snake

Pseudonaja mengdeni

TL 1.2 m

Midbody scales in 17 rows; 19–24 scale rows level with 1st ventral scale; blackish mouth-lining; iris dark with broken reddish orange ring around pupil; rounded snout.

Numerous extremely variable colour forms occur, incl.; plain brown to reddish brown; yellow to bright orange with dark reticulated pattern and glossy black head; prominent broad dark bands sometimes alternating with 3 narrower dark bands within pale interspaces; brown to yellowish brown, with fine reticulated pattern, paler head and broad dark band across neck; black W- or V-shape on neck.

Arid areas, mainly sandy soils with spinifex, in CHC. Also wstn NSW, NT and WA but distributional limits ill-defined pending assessment of museum specimens currently assigned to *P. nuchalis*. DANGEROUSLY VENOMOUS.

Ringed Brown Snake

Pseudonaja modesta

TL 600mm

Midbody scales in 17 rows; blackish mouth-lining; orange-brown iris; widely spaced dark bands; prominent black head and nape markings of juv. retained to adulthood; very small size. Grey to rich reddish brown with up to 12 evenly spaced narrow dark bands

Pseudonaja modesta. Muttaburra.

E. Vanderduys

Pseudonaja nuchalis. Archer River, Qld.

Pseudonaja modesta. Adult without pattern. Eromanga area..

M. Anthony

Pseudonaja nuchalis. Georgetown.

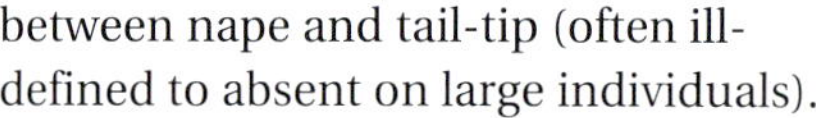

between nape and tail-tip (often ill-defined to absent on large individuals).

Semi-arid woodlands to arid shrublands, spinifex deserts of ML, CHC, MGD, NWH. Also arid zones of all mainland States except Vic. Not normally considered dangerous because of its small size, but like all members of genus should be treated with extreme caution.

Northern Brown Snake

Pseudonaja nuchalis

TL 1.3m

Midbody scales in 17 rows; 23–27 scale rows level with 1st ventral scale; blackish mouth-lining; iris dark with orange ring around pupil; large strap-like rostral scale creating squared snout in dorsal view. Extremely variable, ranging from brown to yellow, often with

M. Anthony

Pseudonaja nuchalis. Almaden.

broad dark bands, and black scales on neck, either scattered or in W- or V-shape. Head and neck may be uniformly dark, or snout may be pale in contrast to darker base of head.

Tropical woodlands of CYP, EIU, GUP and NWH, s. to about Mt Isa and Charters Towers. Also nthn NT and WA but distributional limits ill-defined pending assessment of museum specimens of other spp. currently assigned to *P. nuchalis*. DANGEROUSLY VENOMOUS.

Common Brown Snake; Eastern Brown Snake

Pseudonaja textilis

TL 2.2m

Midbody scales in 17 rows; pink mouth-lining; pale brown iris. Colour ranges through shades of brown to almost black, darkest individuals often having paler heads. Adults can be patternless, or with dark and/or pale flecking, occasionally prominent dark bands. In addition to black head-blotch and neck-band, some juv. have prominent narrow black bands on body. Same litter can contain banded and unbanded hatchlings. Ventral surfaces cream, yellow to orange, with scattered orange to grey blotches.

Dry areas throughout Qld, possibly excl. GUP, parts of CYP. Also all mainland States, estn NG. Most frequently encountered large venomous snake, thriving on edges of towns and cities. Particularly successful in agricultural regions, where tree clearing and introduction of house mice combine to create ideal conditions. Extremely nervous and easily provoked. DANGEROUSLY VENOMOUS.

Pseudonaja textilis. Banded juvenile. Brisbane.

Pseudonaja textilis. Banded adult. Morayfield.

Pseudonaja textilis. Threat posture. Moggill.

Pseudonaja textilis. Mating. East Lynne Stn, Eromanga area.

Genus *SUTA*

11 Aust. spp.; 3 occur in Qld. Relatively depressed heads; eyes with pale or dark irises and vertically elliptic pupils; nasal scale contacting preocular, or separated by contact between prefrontal and upper labials; smooth, glossy scales in 15–19 (rarely 21) midbody rows; single anal and subcaudal scales.

Dry areas throughout Qld. Also dry to well-drained areas in all mainland states. Terrestrial, sheltering in soil cavities, disused termite nests, under rocks, logs and other surface debris. Nocturnal predators of small vertebrates. Livebearers. Large individuals should be treated with extreme caution.

KEY TO *SUTA*

1 Midbody scales in 15 rows 2
Midbody scales in 17 or more rows .... ***suta***

2 Head the colour of body or paler, with dark spots and blotches .... ***punctata***
A solid dark hood on head .... ***dwyeri***

Dwyer's Snake
Suta dwyeri
TL 600mm

Dark iris; nasal scale contacts preocular scale; 15 midbody scale rows; black patch on head. Brown to reddish brown with dark base to each scale, a glossy black head blotch, and cream to brown lips, tip of snout and area in front of eye.

Dry woodlands and outcrops in BB, SEQ, NET and ML. Also NSW and Vic. Predator of small lizards, mainly skinks.

Suta dwyeri. Bendidee SF.

Not regarded as dangerous although a large individual could produce uncomfortable symptoms.

Little Spotted Snake
Suta punctata
TL 523mm

Dark iris; 15 midbody scale rows; spotted head and neck. Yellowish brown to rich reddish brown with darker base to each scale. Head usually paler and yellower, with distinctive dark markings which typically incl. streak

S. Macdonald

Suta punctata. Doomadgee.

from snout through or under eye to base of head, spots on snout and top of head, larger blotches on back of head and sides of neck.

Woodlands, spinifex grasslands of wstn GUP, NWH, MGD, nthn CHC. Also NT, nthn WA. Feeds largely on lizards. Not regarded as dangerous.

Suta suta. Iona Stn.

Myall Snake; Curl Snake
Suta suta
TL 600mm

Pale iris; 19 (rarely 17 or 21) midbody scale rows; dark hood, dark-edged pale stripe on face. Shades of brown to olive, conspicuously darker on top of head (less discernible on dark individuals), with pale brown streak running through eye to temple, dark lower edge contrasting sharply with cream lips.

Widespread in dry areas across Qld, mainly in woodlands, grasslands on heavy and cracking soils. Absent only from SEQ, NET, CQC, WT. Also dry parts of all mainland States. Feeds largely on lizards; frogs, small mammals also taken. Large individuals considered potentially DANGEROUSLY VENOMOUS.

Genus *TROPIDECHIS* Monotypic genus.

Rough-scaled Snake; Clarence River Snake
Tropidechis carinatus
TL 900mm

Keeled scales; deep squarish head; large eye with pale iris and round pupil; matt-textured scales in 23 midbody rows; single anal and subcaudal scales. Shades of brown to olive with irregular darker bands or transversely oriented blotches.

Widely separated pops, in SEQ n. to Fraser Is., and in WT between Mt Spec and Thornton Peak. Also ne. NSW. Wet sclerophyll forests, rainforests, mainly in upland areas; lowland pops occur in parts of Brisbane (e.g. Taringa). Largely terrestrial but very capable climber in thick vegtn. Nocturnal and diurnal; generalist predator of small vertebrates. Livebearer. DANGEROUSLY VENOMOUS.

Tropidechis carinatus. Mt Glorious.

Bandy Bandys
Genus *VERMICELLA*

6 Aust. spp.; 3 occur in Qld. Cylindrical bodies; narrow heads with small eyes and rounded snouts; short tails; smooth, glossy scales in 15 midbody rows; divided anal and subcaudal scales; sharply contrasting black and white rings or bands.

Burrowers, emerging at night, particularly after rain. They feed almost exclusively on blind snakes (Typhlopidae), rarely taking other small snakes. When provoked bandy bandys respond with a display designed to startle, contorting and raising the body into vertically oriented hoops which are realigned as the snake thrashes and pauses. In motion the black and white markings blur to grey, an illusion called 'flicker fusion'. Egglayers. Not regarded as dangerous

KEY TO *VERMICELLA*

1	Ventral scales 260 or more; NWH only	***vermiformis***
	Ventral scales 257 or fewer	2
2	Markings usually form complete rings; white rings sharp-edged, comprising white scales without dark edges; widespread in Qld	***annulata***
	Markings do not extend across ventral surface; white bands with ragged or diffuse edges, comprising dark-edged white scales; nw CYP only	***parscauda***

Common Bandy Bandy
Vermicella annulata

TL 760mm

Relatively robust build; usually fewer than 257 ventral scales. Simply and sharply patterned with 48–147 black and white rings encircling body and tail.

Very broad range of habitats, from cool upland rainforests to sandy deserts and tropical woodlands, in all bioregions with possible exception of CHC.

Vermicella annulata. Mt Nebo.

L. Dibben

Vermicella parscauda. Weipa.

S. Eipper

Vermicella vermiformis. Mt Isa area.

Cape York Bandy Bandy

Vermicella parscauda

TL 388mm

Relatively slender build; 213–230 ventral scales. Prominently marked with 110–184 black and narrower white bands, forming complete rings only on posterior body and tail. The white bands, comprising dark-edged pale scales, have diffuse or ragged edges. Ventral surfaces dark or mottled.

Endemic. Woodlands on heavy red soils between Weipa, Mapoon and Bamaga areas, CYP.

Centralian Bandy Bandy

Vermicella vermiformis

TL 525mm

Slender build; 260–280 ventral scales. Prominently marked with 81–90 black and white rings encircling body and tail.

Woodlands, often on stony soils in vicinity of Mt Isa, NWH. Also known from widely separated localities in nthn and sthn NT.

SEA SNAKES

R. Somaweera

The Yellow-bellied Sea Snake (*Hydrophis platurus*) extends along the length of Queensland's coastline. Most records are beach-washed specimens, but this is a truly pelagic species, occurring over vast tracts of the earth's oceans, often well away from the reefs and shallower seas inhabited by other sea snakes.

Family **Elapidae**; Subfamily **Hydrophiinae**

Fully aquatic, venomous snakes with fixed fangs at the front of the mouth, dorsally placed, valvular nostrils and laterally flattened, paddle-shaped tails. Ventral scales are often greatly reduced, sometimes no larger than adjacent scales, allowing the body to become laterally compressed for easy passage through water.

Most species feed on fish, immobilising them with powerful venom. Some generalists take a variety of prey and others specialise on eels. Two Queensland species, *Emydocephalus annulatus* and *Aipysurus mosaicus* have reduced venom apparatus and eat only fish eggs.

All sea snakes are live bearers, so with feeding and reproduction taking place in the sea, they need never venture onto land. Those encountered on beaches are normally stranded due to illness or injury and can be carried outside of their usual distribution by ocean currents. In the absence of enlarged ventral scales, they are not as mobile as terrestrial elapids. However some species with broad ventral scales, such as *Aipysurus* spp., can move on land, while *Hydrelaps darwiniensis*, and *Parahydrophis mertoni* occupy intertidal mudflats.

Sea snakes reside in all Queensland waters, with diversity highest in the tropics. Occasionally northern species reach southern beaches, and a few not normally present in Australia are rarely recorded in Torres Strait. They occur throughout the Great Barrier Reef yet surprisingly, given its immense size and biological diversity, the reef harbours no endemic species. Anecdotal and scientific evidence suggests that some species, such as olive sea snakes, may be declining. Excepting the two egg-eating species, all sea snakes should be regarded as capable of a fatal envenomation if handled or harassed.

KEY TO GENERA

1	Ventral scales about 3 times wider than adjacent body scales	2
	Ventral scales range from no larger than adjacent body scales to no more than twice as wide	***Hydrophis***
2(1)	Only 3 upper labial scales **(a)**	***Emydocephalus***
	More than 3 upper labial scales	3
3(2)	A distinct preocular scale present	4
	No distinct preocular scale	***Hydrelaps***
4(3)	Midbody scales 25 or fewer	***Aipysurus***
	Midbody scales more than 30	***Parahydrophis***

a

Genus *AIPYSURUS*

8 spp. in Aust.; 3 occur in Qld. Large ventral scales at least 3 times broader than adj. body scales, each usually with a keel and posterior notch on midline; at least 6 upper labial scales; head scales either fragmented or large and symmetrical; body scales strongly to weakly overlapping, often with tubercles on males, with vertebral row usually enlarged.

Most live in seas shallower than 50 m, between South-East Asia and wstn Pacific. They capture fish and invertebrates in benthic habitats ranging from soft bottoms and seagrass beds, to shoals, coral and rocky reefs. One is a fish-egg specialist.

KEY TO *AIPYSURUS*

1	All head shields large and symmetrical	***mosaicus***
	At least some head shields fragmented	2
2	All head shields greatly fragmented **(a)**; no larger than adjacent neck scales	***duboisii***
	Head shield fragmentation restricted to parietals; remainder large and symmetrical	***laevis***

a

Dubois' Sea Snake

Aipysurus duboisii

TL 1.1m

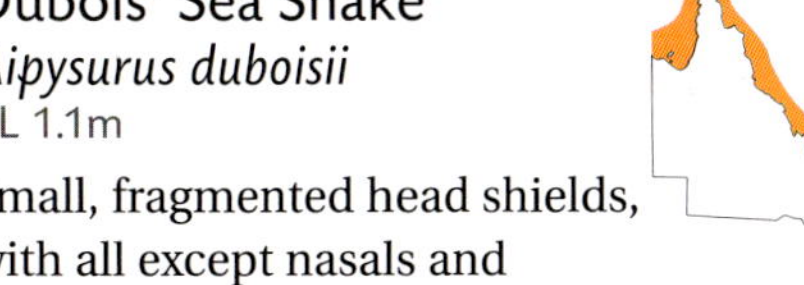

Small, fragmented head shields, with all except nasals and rostral no larger than scales on neck; mid-ventral keels very weak and notches shallow; body scales smooth to weakly keeled or tuberculate, overlapping, in 19 rows at midbody. Extremely variable, but

H. Cogger

Aipysurus duboisii. Ashmore Reef, WA.

typically purplish brown with pale-edged scales, often with cream bands tending to taper to form lateral triangles.

All coastal waters, from shallow reef flats to depths to 50m. Also nthn Aust. to NG and New Caledonia, and s. to NSW. DANGEROUSLY VENOMOUS.

Aipysurus laevis. North-western WA. H. Cogger

Olive Sea Snake

Aipysurus laevis

TL 1.7m

Most head shields large and symmetrical, but some partly fragmented on base of head; mid-ventral keels moderate and notches present; scales smooth and overlapping in 21–25 midbody rows; very large size, robust build. Extremely variable, from golden brown or olive, with or without scattered dark scales, to purplish brown merging evenly with pale ventral surfaces, or with scattered sharply contrasting cream-spotted scales.

All coastal waters, particularly coral reefs. Also nthn NSW to nthn Aust., NG and New Caledonia. Generalist fish predator. DANGEROUSLY VENOMOUS.

G. Schmida

Aipysurus mosaicus. Northeastern Qld.

Mosaic Sea Snake

Aipysurus mosaicus

TL 110mm

Large, symmetrically arranged head shields; mid-ventral keels weak; ventral scales with no posterior notch; body scales smooth, just overlapping in 17 rows at midbody. Cream, salmon or yellowish brown to golden brown, usually with many scales dark-edged, and numerous irregular, often broken dark bands tending to form staggered half bands on either side.

All coastal waters s. to SEQ, particularly turbid waters about 30–50m deep. Also nthn Aust., NG, South-East Asia. Eats only fish eggs. Reluctant to bite. Not considered dangerous.

Genus *EMYDOCEPHALUS*

3 spp.; 1 in South China Sea and 2 in Aust. 1 sp. occurs in Qld.

Turtle-headed Sea Snake

Emydocephalus annulatus

TL 950mm

Unique turtle-like profile created by short, blunt head and plate-like upper lips, composed of only 3 upper labial scales; conical spur on rostral scale; ventral scales much wider than adj. scales. Black, dark brown or grey, with or without prominent cream bands.

Shallow waters on coral reefs. Recorded from SEQ, CQC (Swain Reefs) and probably occurs along other parts of Qld coast but appears absent from Gulf of Carpentaria. Also New Caledonia, across nthn Aust. to Timor Sea. Eats only fish eggs, scraped from substrate with plate-like upper lip. Mainly diurnal. Harmless.

H. Cogger

Emydocephalus annulatus. Ashmore Reef, WA.

Genus *HYDRELAPS* Monotypic genus.

Black-ringed Mangrove Snake

Hydrelaps darwiniensis

TL 530mm

Broad ventral scales about 3 times width of adj. body scales; no preocular scales; large symmetrical head shields; smooth overlapping body scales in 25–30 midbody rows; bold dark rings. Cream to pale grey with 30 or more sharp, dark rings, widest and often enclosing pale blotch on back.

Coastal waters, mudflats of Gulf of Carpentaria. Also NT, WA, sthn NG. Forages for fish in shallow water along rising and falling tide-lines. Not considered dangerous.

P. Horner.

Hydrelaps darwiniensis. Bing Bong, NT.

Genus *HYDROPHIS*

35+ spp worldwide; 17 in Aust.; 13 occur in Qld. Extremely ecologically diverse genus, rapidly speciating, with extreme variability among members. Ventral scales small; indiscernible, to scarcely larger or up to twice as large as adj. scales, entire or divided into 2 rows. Head shields large, non-fragmented symmetrical plates (fragmented on one Aust. sp.). Spp. vary from relatively unspecialised, generalist fish predators, to snakes with robust hindbodies, extremely thin forebodies and small heads, designed to penetrate narrow burrows for eels. Incl. inhabitants of clear or turbid, deep or shallow waters. One cosmopolitan sp. is pelagic.

Hydrophis can be extremely difficult to identify to species level, due to a lack of accurate, lineage-wide diagnostic features, requiring genetic confirmation of species identity along with examination of internal features such as the skeleton and heart position. This key is adaptedfrom Rasmussen *et al*, 2014. It relies in part on a combination of characters including scales around the neck and midbody. These are minimum and maximum counts respectively. The minimum neck count is achieved by taking about four counts and selecting the lowest. These are at one and half, two, two and a half, and three head lengths behind the head. Maximum body scale count is posterior to midbody, and three to four counts between midbody and vent are recommended to locate the highest number.

KEY TO *HYDROPHIS*

a **b** **c**

1 Mental scale very long and dagger-shaped **(a)**........ ***zweifeli***
Mental scale triangular, not much longer than broad **(b)**.... 2

2(1) Ventral scales in an overlapping divided series........ ***stokesii***
Ventral scales entire throughout all or most of their length.. 3

3(2) Head scales fragmented; raised posterior spines on some head shield, particularly above eyes **(c)** .... ***peronii***
Head scales large symmetrical shields with no raised projecting spines 4

4(3) Sharply contrasting black back and yellow belly..... ***platurus***
Not as above... 5

5(4) Body scales non-overlapping; ventral scales unrecognizable posteriorly................................. ***curtus***
Body scales overlapping and ventral scales recognizable 6

6(5) Fewer than 275 ventral scales and less than 30 scale rows around neck.. 7
Not as abovc.. 8

7(6) 2–3 series of spots or dark-edged markings between bands and on sides; head very small and body compressed posteriorly.. ***macdowelli***
Not as above.. 8

8(6,7) Fewer than 275 ventral scales and fewer than 44 midbody scale rows.. 9
Not as above.. 10

9(7) Fewer than 38 bands on body and small dark spots on flanks .... ***major***
Not as above.. 10

10(8,9) Fewer than 37 midbody scale rows; fewer than 320 ventral scales and more than 45 bands on body and tail........ ***donaldi***
Not as above........ 11

11(10) More than 345 ventral scales and spots on flanks and between bands........ ***elegans***
Not as above........ 12

12 (11) More than 30 scale rows around neck, pale ground colour with ocelli on sides........ ***ocellatus***
Not as above........ 13

13(12) More than 30 scale rows around neck; ground colour not yellow; more than 12 maxillary teeth behind fang........ ***caerulescens***
Not as above........ 14

14 (13) Head small and neck elongated; body more than 3 times as broad as head; white ring around eye.. ***kingii***
Not as above........ 15

15 (14) More than 40 bands on body and tail........ ***pacificus***
Less than 40 bands on body and tail........ ***coggeri***

H. Cogger

Hydrophis caerulescens. Preserved holotype. Indian Ocean.

H. Cogger

Hydrophis coggeri. Ashmore Reef, WA.

Dwarf Sea Snake

Hydrophis caerulescens
TL 1m

Strongly keeled, overlapping scales in 37–39 midbody rows. Ventral scales scarcely wider than adj. scales; small head, slender forebody and relatively robust, laterally compressed hindbody. Bluish grey with dark head and 41–43 blackish rings, tapering towards belly and sometimes disrupted and alternating on vertebral line. Tail black with bands, or with pale blotches along dorsal edge.

Recorded from se. Gulf of Carpentaria and tentatively identified from estn QLD. Also South-East Asia to Pakistan. Captures eels and gobies on sea floor. DANGEROUSLY VENOMOUS.

Cogger's Sea Snake

Hydrophis coggeri
TL 1.3m

Scales smooth, just overlapping anteriorly and arranged side by side posteriorly in 29–34 midbody rows; ventral scales scarcely wider than adj. scales; small head, slender forebody and relatively robust, laterally compressed hind body. Adults olive grey with 28–40 dark rings widest dorsally and ventrally. Juv. much more prominently marked with pale yellow and black rings, and black head with conspicuous yellow marks on snout and behind eyes.

Recorded in waters off wstn CYP. Also off nw. WA and wstn Pacific. DANGEROUSLY VENOMOUS.

Hydrophis curtus. Unity Reef.

Spine-bellied Sea Snake
Hydrophis curtus
TL 1m

Hexagonal to squarish, non-overlapping body scales, largest on lower flanks, in 23–45 midbody rows; often prominent tubercles or spines on flanks; large symmetrical head shields; greatly reduced ventral scales; large head, robust build. Olive to brown above, cream below, colours meeting along deeply indented angular junction which can almost break upper colour into bands.

Coastal waters s. to CQC. Also nthn Aust. to Arabian Gulf. Feeds on variety of fish. Has caused fatalities. DANGEROUSLY VENOMOUS.

K. Ukewela

Hydrophis donaldi. Weipa area.

Rough-scaled Sea Snake
Hydrophis donaldi
TL 878mm

Strongly spinose scales. Moderately robust with midbody scales in 33–35 rows. Yellowish brown with 47–56 darker brown bands on body and tail. Juv. brighter with dark head and turquoise interspaces between bands.

Endemic. Known only from Weipa area, in estuarine habitats less than 10 m deep over shale, mud and sea-grass where mouths of Mission River and Hey Ck enter Albatross Bay.

Hydrophis elegans. Surfers Paradise.

Elegant Sea Snake
Hydrophis elegans
TL 2m

Scales smooth or with short keels, overlapping in 37–49 midbody rows; ventral scales barely wider than adj. scales; slender neck and forebody and very thick, laterally compressed hindbody; large size. Brown to olive with 35–55 broad blackish bands or blotches across back and normally 2 rows of blotches along flanks, a midlateral row placed within the pale interspaces and a lower lateral row aligned with dorsal markings.

All coastal waters. Also central coastal NSW to nthn Aust., NG. Feeds on eels. DANGEROUSLY VENOMOUS.

H. Cogger

Hydrophis kingii. Ashmore Reef, WA.

G. Gow

Hydrophis macdowelli. Arafura Sea.

Spectacled Sea Snake
Hydrophis kingii
TL 1.9m

Ventral scales about twice as wide as adj. scales, each with one keel; small head and slender forebody; scales keeled and easily overlapping on back, but smooth and just overlapping on sides, in 34–39 rows at thickest part of body; anterior chin shields large and easily contacting mental groove. Head black with white ring around eye, narrow white bar around neck. Body grey above, paler on midlateral and ventral surfaces, with numerous dark grey blotches extending onto lower flanks. Ventral scales black.

All coastal waters. Also central coastal NSW to nthn Aust., sthn NG. DANGEROUSLY VENOMOUS.

Small-headed Sea Snake
Hydrophis macdowelli
TL 800mm

Scales just overlapping in 35–42 midbody rows and keeled on neck; extremely small head, very thin forebody and thick, deeply laterally compressed hindbody. Cream with dark head, dark triangular dorsal blotches reaching about one-third down flanks, 3 lateral series of dark blotches, 4–5 pale grey bands on tail.

Coastal waters between SEQ and Gulf of Carpentaria, favouring turbid estuaries and inshore waters. Also nthn Aust to New Caledonia. Eats eels, other long-bodied fish. Probably DANGEROUSLY VENOMOUS.

Hydrophis major. Gulf of Carpentaria.

Greater Sea Snake
Hydrophis major
TL 1.6m

Ventral scales slightly wider than adj. scales, each with 2 keels; short thick head and robust body; scales bluntly keeled and just overlapping, in 33–43 rows at thickest part of body; anterior chin shields just contacting or excluded from mental groove by long 1st lower labial scales. Pale grey to olive with 24–30 bold (juv.) to diffuse (adult) broad dark bars across back. Narrow dark bars within each pale interspace align with series of dark spots along flanks.

All deep, turbid coastal waters, commonly extending s. to SEQ. Also central coastal NSW to nthn Aust., sthn NG. DANGEROUSLY VENOMOUS.

A. Rasmussen

Hydrophis ocellatus. Off Broome, WA.

Ocellated Sea Snake

Hydrophis ocellatus
TL 1.5m

Scales overlapping in 39–59 midbody rows; robust body of roughly uniform thickness; ventral scales about twice as wide as adj. scales. Grey with 30–60 broad dark bands or dorsal blotches and large spots on flanks. With age pattern reduces to virtual absence.

Reported to be widespread along coast. Also NSW to nthn Aust., wstn Pacific to Persian Gulf. Feeds on variety of fish. DANGEROUSLY VENOMOUS.

Large-headed Sea Snake

Hydrophis pacificus
TL 1.4m

Scales overlapping in 39–49 midbody rows; large head; slender forebody and robust, laterally compressed hindbody. Head black with yellow spots on snout and behind eyes. Body dark grey above, becoming abruptly paler at mid-flanks with 49–72 dark bands, often offset along midline. Pattern sharpest on juv.

Gulf of Carpentaria. Also Arafura Sea, sthn NG. DANGEROUSLY VENOMOUS.

H. Cogger

Hydrophis pacificus. Preserved holotype. New Britain, PNG.

H. Cogger

Hydrophis peronii. Ashmore Reef, WA.

Horned Sea Snake

Hydrophis peronii
TL 1.23m

Fragmented head shields; scales above eyes raised to form horn-like spines; ventral scales no larger than adj. scales; slender forebody, robust hindbody; scales weakly overlapping, each with one keel, often dark, in 27–29 rows at thickest part of body. Brownish white, with or without darker bands that taper on flanks.

All coastal waters s. to SEQ, often assoc. with sandy areas near coral. Also nthn Aust. to New Caledonia. Largely nocturnal predator of gobies. DANGEROUSLY VENOMOUS.

E. Vanderduys

Hydrophis platurus. Moreton Is.

Yellow-bellied Sea Snake

Hydrophis platurus

TL 1m

Non-overlapping scales in 47–69 midbody rows; large symmetrical head shields; very small ventral scales; sharply contrasting dorsal and ventral colours, black above and yellow below. Tail yellow with black spots or bars.

All Qld waters. Also around Aust. and all global tropical, subtropical and sometimes temperate seas except Atlantic, primarily in open ocean rather than over continental shelves. Eats wide variety of fish, captured by a 'float and wait' strategy as snake drifts along slick lines where currents merge. DANGEROUSLY VENOMOUS.

P. Horner

Hydrophis stokesii. Darwin Harbour, NT.

Stokes' Sea Snake

Hydrophis stokesii

TL 2m

Head shields large and symmetrical; scales overlapping in 46–63 rows at midbody; ventral scales scarcely larger than adj. scales, arranged in 2 overlapping rows forming obvious ventral keel; extremely large size, robust build (most massive sea snake), incl. large head and thick neck; longest fangs of Aust., sea snakes (4.2mm). Cream to brown or yellowish brown with large dark dorsal blotches alternating with narrow bands. Pattern normally obvious on juv., tending to fade with age. Adults often patternless.

All coastal waters. Also nthn NSW, across nthn Aust. to Pakistan and Arabian Gulf. Generalist fish predator. DANGEROUSLY VENOMOUS.

Beaked Sea Snake

Hydrophis zweifelli

TL 1.2m

Enlarged symmetrical head shields; extremely long, dagger-shaped mental scale; body scales overlapping, each with short low keel, in 49–66 rows at midbody; ventral scales scarcely

wider than adj. scales; rear of body strongly laterally compressed. Grey with diffuse-edged dark bars across back, dark grey head.

Scattered localities between SEQ and wstn CYP, mainly in shallow bays and estuaries. Also NG. Eats fish, particularly catfish and pufferfish. DANGEROUSLY VENOMOUS.

L. Dibben

Hydrophis zweifelli. Weipa area.

Genus *PARAHYDROPHIS* Monotypic genus.

Northern Mangrove Sea Snake
Parahydrophis mertoni
TL 500mm

Relatively wide ventral scales, about 3 times as wide as adj. body scales; large symmetrical head shields; body scales smooth and overlapping in 36–39 midbody rows. Bluish grey to greyish brown with about 40–50 darker bands or rings, often enclosing pale mid-dorsal blotch.

Coastal waters of Gulf of Carpentaria. Also nthn NT, Aru Is. off wstn NG. Probably assoc. with intertidal shallows.

H. Cogger

Parahydrophis mertoni. Darwin, NT.

SEA KRAITS

Laticauda colubrina. Morovo Lagoon, Solomon Islands.

Family **Elapidae**; Subfamily **Laticaudinae**

Amphibious marine snakes with short, fixed fangs at the front of the mouth, large symmetrical head shields, nasal scales separated by internasals, laterally positioned nostrils, moderately fragmented lower labial scales, laterally flattened, paddle-shaped tails, smooth overlapping body scales, broad expanded ventral scales and a pattern of sharply contrasting rings around body. All 4 members in subfamily placed in the genus *Laticauda.*

Sea kraits feature a mix of terrestrial and aquatic modifications, placing them somewhere between the marine and terrestrial elapid subfamilies. They forage for eels in the sea, but shelter and lay eggs on land, usually on rocky islets.

None resides in Qld, but they are common in tropical seas to the northeast and northwest. Only one species, *Laticauda colubrina*, has been recorded as waifs carried by currents. Another, *Laticauda laticaudata*, is often listed but it seems there are no reliable records. They are believed to have strong venom but are placid and reluctant to bite. DANGEROUSLY VENOMOUS.

Yellow-lipped Sea-krait

Laticauda colubrina

TL 1.4m

Yellow snout and upper lips; 21–25 midbody scale rows.

Blue to blue-grey, with 20–65 prominent black rings. Top of head black.

Occasional strays recorded from nthn Torres Strait to reefs off Mackay, and from NSW. Also sw. Pacific to India and Sri Lanka.

SELECTED READING

Amey, A. P., Couper, P. J and Worthington Wilmer, J., 2019. A new species of *Lerista* Bell, 1833 (Reptilia: Scincidae) from Cape York Peninsula, Queensland, belonging to the *Lerista allanae* clade but strongly disjunct from other members of the clade. *Zootaxa* 4613 (1): 161–171.

Amey, A. P., Couper, P. J and Worthington Wilmer, J., 2019. Two new species of *Lerista* Bell, 1833 (Reptilia: Scincidae) from north Queensland populations formerly assigned to *Lerista storri* Greer, McDonald and Lawrie, 1983. *Zootaxa* 4577 (3): 473–493.

Brown, D., Worthington Wilmer, J. and Macdonald, S. (2012). 'A revision of *Strophurus taenicauda* (Squamata: Diplodactylidae) with a description of two new subspecies from central Queensland and a southerly range extension.' *Zootaxa,* 3243: 1–28.

Cann, J. and Sadlier, R. 2017. *Freshwater Turtles of Australia.* CSIRO Publishing. Clayton South, Vic.

Chapple, D. G., Tingley, R., Mitchell, N. J., Macdonald, S. L., Keogh, J. S., Shea, G. M., Bowles, P., Cox, N. A. and Woinarski, J. Z. 2019. *The Action Plan for Australian Lizards and Snakes 2017.* CSIRO Publishing, Clayton South, Vic.

Cogger, H. G., 2014. *Reptiles and Amphibians of Australia.* Seventh edition. CSIRO Publishing. Collingwood, Vic.

Cogger, H. G., Ford, H., Johnson, C., Holman, J. and Butler, D., 2003. *Impacts of Land Clearing on Australian Wildlife in Queensland.* World Wildlife Fund, Australia.

Couper, P. J., Amey, A. P. and Kutt, A. S., 2002. 'A new species of *Ctenotus* (Scincidae) from Central Queensland'. *Memoirs of the Queensland Museum* 48(1): 85–91.

Couper, P. J., Covacevich, J. A. and Wilson, S. K., 1998. 'Two new species of *Ramphotyphlops* (Squamata: Typhlopidae) from Queensland'. *Memoirs of the Queensland Museum* 42(2): 459–64.

Couper, P. J. and Hoskin, C. J., 2013. 'Two new subspecies of the leaf-tailed gecko *Phyllurus ossa* (Lacertilia: Carphodactylidae) from mid-eastern Queensland, Australia.' *Zootaxa,* 3664 (4): 537–553.

Couper, P. J., Hoskin, C. J., Potter, S., Bragg, J. G. and Moritz, C., 2018. A new genus to accommodate three skinks currently assigned to *Proablepharus* (Lacertilia: Scincidae). *Mem. Qld. Mus.– Nature.* 60: 227–231.

Couper, P. J., Keim, D. and Hoskin, C. J., 2007. 'A new velvet gecko (Gekkonidae: *Oedura*) from south-east Queensland, Australia.' *Zootaxa* 1587: 27–41.

Couper, P. J. and Keim, L. D., 1998. 'Two new species of *Saproscincus* (Reptilia: Scincidae) from Queensland'. *Memoirs of the Queensland Museum* 42(2): 465–73.

Couper, P. J., Limpus, C. J., McDonald, K. R. and Amey, A. P., 2010. 'A new species of *Proablepharus* (Scincidae: Lygosominae) from Mt Surprise, north-eastern Queensland, Australia.' *Zootaxa* 2433: 62–68.

Couper, P., Melville, J., Emmott, A. and Chapple, S. N. J., 2012. 'A new species of *Diporiphora* from the Goneaway Tablelands of Western Queensland.' *Zootaxa,* 3556: 39–54.

Couper, P. J., Peck, S. R., Emery, J. and Keogh, J. S., 2016. Two snakes from eastern Australia (Serpentes: Elapidae); a revised concept of *Antaioserpens warro* (De Vis) and a redescription of *A. albiceps* (Boulenger). *Zootaxa* 4097 (3): 396–408.

Couper, P. J., Schneider, C. J. and Covacevich, J. A., 1997. 'Two new species of *Saltuarius* (Lacertilia: Gekkonidae) from granite-based, open forests of eastern Australia'. *Memoirs of the Queensland Museum* 42(1): 91–96.

Couper, P. J., Schneider, C. J., Hoskin, C. J. and Covacevich, J. A., 2000. 'Australian leaf-tailed geckos: Phylogeny, a new genus, two new species and other data'. *Memoirs of the Queensland Museum* 45(2): 253–65.

Covacevich, J. A, Couper, P. J. and McDonald, K. R., 1998. 'Reptile diversity at risk in the Brigalow Belt, Queensland'. *Memoirs of the Queensland Museum* 42(2): 475–86.

Derez, C. M., Arbuckle, K., Ruan, Z., Xie, B., Huang, Y., Dibben, L., Shi, Q., Zonk, F. and Fry, B. G. 2018. A new species of bandy-bandy (*Vermicella*: Serpentes: Elapidae) from the Weipa region, Cape York, Australia. *Zootaxa* 4446 (1): 001–012.

Donnellan, S. C., Couper, P. J., Saint, K. M. and Wheaton, L., 2009. 'Systematics of the *Carlia 'fusca'* complex (Reptilia: Scincidae) from northern Australia.' *Zootaxa,* 2227: 1–31.

Eastwood, J. A., Doughty, P., Hutchinson, M, N and Pepper, M., 2020. Revision of *Lucasium stenodactylus* (Boulenger, 1896; Squamata: Diplodactylidae) with the resurrection of *L. woodwardi* (Fry, 1914) and the description of a new species from south-central Australia. *Rec. West Aust. Mus.* 35: 63–86.

Edwards, D. L. and Melville, J., 2011. 'Extensive phylogeographic and morphological diversity in *Diporiphora nobbi* (Agamidae) leads to a taxonomic review and a new species description.' *Journal of Herpetology* 45 (4): 530–546.

Ehmann, H., 1992. *Encyclopedia of Australian Animals: Reptiles.* Collins Angus and Robertson, Sydney.

Eipper, S. & Eipper, T. E. 2019. *A Naturalist's Guide to the Snakes of Australia.* Australian Geographic. John Beaufoy Publishing.

Emmott, A. and Wilson, S. G., 2009. *Snakes of Western Queensland. A Field Guide.* Desert Channels Queensland, Longreach, Qld.

Esquerre, D., Donnellan, S. C., Pavon-Vazquez, C. J., Fenker, J. and Keogh, J. S., 2021. Phylogeny, historical demography and systematics of the world's smallest pythons (Pythonidae, *Antaresia*). *Molecular Physics and Evolution* 161: 107–181.

Greer, A. E., 1989. *The Biology and Evolution of Australian Lizards.* Surrey Beatty and Sons, Chipping Norton, NSW.

Greer, A. E., 1991. 'Two new species of *Menetia* from northeastern Queensland with comments in the generic diagnoses of *Lygisaurus* and *Menetia*'. *Journal of Herpetology* 25(3): 268–72.

Greer, A. E., 1997. *The Biology and Evolution of Australian Snakes.* Surrey Beatty and Sons, Chipping Norton, NSW.

Heatwole, H., 1999. *Sea Snakes.* Australian Natural History Series. UNSW Press, Sydney.

Hedges, S. B., Marion, A. B., Lipp, K. M., Marin, J. and Vidal, N., 2014. 'A taxonomic framework for typhlopid snakes from the Caribbean and other regions (Reptilia, Squamata).' *Caribbean Herpetology,* 49: 1–61.

Horner, P., 2007. 'Systematics of the snake-eyed skinks, *Cryptoblepharus* Wiegmann (Reptilia: Squamata: Scincidae) – an Australian-based review.' *The Beagle, Records of the Museums and Art Galleries of the Northern Territory, Supplement 3.*

Hoskin, C. J., 2013. 'A new skink (Scincidae: *Saproscincus*) from rocky forest habitat on Cape Melville, north-east Australia.' *Zootaxa,* 3722: 385–395.

Hoskin, C. J., 2014. 'A new skink (Scincidae: *Carlia*) from the rainforest uplands of Cape Melville, north-east Australia.' *Zootaxa,* 3869 (3): 224–236.

Hoskin, C. J., 2019. Description of three new velvet geckos (Diplodactylidae: *Oedura*) from inland eastern Australia, and redescription of *Oedura monilis* De Vis. *Zootaxa* 4683 (3): 242–270.

Hoskin, C. J., Bertola, L. V. and Higgie M., 2019. A new species of *Phyllurus* leaf-tailed gecko (Lacertilia: Carphodactylidae) from The Pinnacles, north-east Australia. *Zootaxa* 4576 (1): 127–139.

Hoskin, C. J. and Couper, P. J., 2013. 'A spectacular new leaf-tailed gecko (Carphodactylidae: *Saltuarius*) from the Melville Range, north-east Australia.' *Zootaxa,* 3717 (4): 543–558.

Hoskin, C. J. and Couper, P. J., 2014. 'Two new skinks (Scincidae: *Glaphyromorphus*) from rainforest habitats in north-eastern Australia.' *Zootaxa,* 3869 (1): 001–016.

Hoskin, C. J., Couper, P. J. and Schneider, C. J., 2003. 'A new species of *Phyllurus* (Lacertilia: Gekkonidae) with a revised phylogeny and key for the Australian leaf-tailed geckos'. *Australian Journal of Zoology* 51: 153–64.

Hoskin, C. J. and Higgie, M., 2008. 'A new species of velvet gecko (Diplodactylidae: *Oedura*) from north-east Queensland, Australia.' *Zootaxa* 1788: 21–36.

Hutchinson, M. N., Couper, P., Amey, A. and Worthington Wilmer, J., 2012. Diversity and systematics of limbless skinks from eastern Australia and the skeletal changes that accompany substrate swimming body form. *J. Herpet.* 55(4): 361–384.

Ingram, G. J. and Raven, R. J., 1991. *An Atlas of Queensland's Frogs, Reptiles, Birds and Mammals.* Board of Trustees of the Queensland Museum, Brisbane.

James, B. H., Donnellan, S. C. and Hutchinson, M., 2001. 'Taxonomic revision of the Australian lizard *Pygopus nigriceps* (Squamata: Gekkonoidea)'. *Records of the South Australian Museum* 34(1): 37–52.

McCord, W. P. and Thompson, S. A., 2002. 'A new species of *Chelodina* (Testudines: Pleurodira: Chelidae) from northern Australia'. *Journal of Herpetology* 36(2): 255–67.

Mecke, S., Doughty, P. and Donnellan, S. C. (2013). 'Redescription of *Eremiascincus fasciolatus* (Gunther, 1867) (Reptilia: Squamata: Scincidae) with clarification of its synomyms and the description of a new species.' *Zootaxa,* 3701 (5): 473–517.

Melville, J., Ritchie, E. G., Chapple, S. N. J., Glor, R. and Schulte, J. 2018. Diversity in Australia's tropical savannas: An integrative taxonomic revision of agamid lizards from the genera *Amphibolurus* and *Lophognathus* (Lacertilia: Agamidae). *Mem. Mus. Vic.* 77: 41–61.

Melville, J., Smith, K., Hobson, R. and Shoo, L., 2014. 'The role of integrative taxonomy in the conservation management of cryptic species: The taxonomic status of endangered earless dragons (Agamidae: *Tympanocryptis*) in the grasslands of Queensland, Australia.' *Plos One,* 9 (7) e101847:1–13.

Melville, J., Smith, K., Horner, P. and Doughty, P., 2019. Taxonomic revision of dragon lizards in the genus *Diporiphora* (Reptilia: Agamidae) from the Australian monsoonal tropics. *Mem. Mus. Vic.* 78: 23–55.

Melville, J. and Wilson, S. K, 2019. *Dragon Lizards of Australia. Evolution, Ecology and a Comprehensive Field Guide.* Museums Victoria Publishing.

Murphy, J. C. 2007. *Homalopsid snakes – Evolution in the mud.* Kreiger Publishing Company, Malabar, Florida, USA.

Oliver, P.M., Couper, P and Amey, A., 2010. 'A new species of *Pygopus* (Pygopodidae; Gekkota; Squamata) from north-eastern Queensland.' *Zootaxa,* 2578: 47–61.

Oliver, P.M., Bauer, A. M., Greenbaum, E., Jackman, T. and Hobbie, T.,2012. 'Molecular phylogenetics of the arboreal Australian gecko genus *Oedura* Gray 1842. (Gekkota: Diplodactylidae): Another plesiomorphic grade?' *Molecular Phylogenetics and Evolution* 63 (2): 255–264.

Oliver, P. M. and Bauer, A. M., 2011. 'Systematics and evolution of the Australian knob-tailed geckos (*Nephrurus,* Carphodactylidae, Gekkota): Plesiomorphic grades and biome shifts through the Miocene'. *Molecular Phylogenetics and Evolution* 59 (3): 664–674.

Oliver. P. M., Prasetya, A. M., Tedeschi, L. G., Fenker, J., Ellis, R. J., Doughty, P. and Moritz, C., 2020. Crypsis and convergence: Integrative taxonomic revision of the *Gehyra australis* group (Squamata: Gekkonidae) from northern Australia. *PeerJ* 8:e7971. DOI 10.7717/peerj.7971.

Pepper, M., Doughty, P. Hutchinson, M. N. and Keogh, J. S., 2011. 'Ancient drainages divide cryptic species in Australia's arid zone: Morphological and multi-gene evidence for four new species of Beaked Geckos (*Rhynchoedura*).' *Molecular Phylogenetics and Evolution.* 61: 810–822.

Pianka, E. R., King, D. R. and King, R. A. (*Eds*) 2004. *Varanoid lizards of the world.* Indiana University Press, USA.

Rasmussen, A. R., Sanders, K. L., Guinea, M. L. and Amey, A. P, 2014. 'Sea snakes in Australian waters (Serpentes: Subfamilies Hydrophiinae and Laticaudinae) – a review with an updated identification key.' *Zootaxa,* 3869 (4): 351–371.

Ross, C. A. (ed.), 1989. *Crocodiles and Alligators.* Golden Press, Sydney.

Sadlier, R. A., Couper, P. J., Colgan, D. J., Vanderduys, E. and Rickard, E., 2005. 'A new species of scincid lizard, *Saproscincus eungellensis,* from mid-eastern Queensland'. *Memoirs of the Queensland Museum.* 51(2): 550–571.

Sanders, K. L., Lee, M. S. Y., Mumpuni, Bertozzi, T. and Rasmussen, A. R., 2013. 'Multolocus phylogeny and recent rapid radiation of the viviparous sea snakes (Elapidae: Hydrophiinae).' *Molecular Phylogenetics and Evolution* 66 (2013): 575–591.

Sattler, P. and Williams, R. (eds), 1999. *The Conservation Status of Queensland's Bioregional Ecosystems.* Environmental Protection Agency, Brisbane.

Shea, G., Couper, P., Worthington Wilmer, J. and Amey, A. 2011. 'Revision of the genus *Cyrtodactylus* Gray, 1827 (Squamata: Gekkonidae) in Australia.' *Zootaxa,* 3146: 1–63.

Shea, G.M. and Scanlon, J. D., 2007. 'Revision of the small tropical whipsnakes previously referred to *Demansia olivacea* (Gray, 1842) and *Demansia torquata* (Gunther, 1862) (Squamata: Elapidae).' *Records of the Australian Museum* (2007) 59 (2): 117–142.

Shine, R., 1991. *Australian Snakes: A Natural History.* Reed New Holland, Sydney.

Shoo, L. P., Rose, R., Doughty, P., Austin, J. J. and Melville, J. 2008. 'Diversification patterns of pebble-mimic dragons are consistent with historical disruption of important habitat corridors in arid Australia.' *Molecular Phylogenetics and Evolution* 48 (2008): 528–542.

Singhal, S., Hoskin, C. J., Couper, P., Potter, S. and Moritz, C., 2018. A framework for resolving cryptic species: a case study from lizards in the Australian Wet Tropics. *Systematic Biology* 67 (6), 1061–1075.

Skinner, A., 2008. 'A multivariate morphometric analysis and systematic review of *Pseudonaja* (Serpentes, Elapidae, Hydrophiinae).' *Zoological Journal of the Linnean Society of London.*

Smith, W. J. S., Osborne, W. S., Donnellan, S. C. and Cooper, P. D., 1999. 'The systematic status of earless dragon lizards, *Tympanocryptis* (Reptilia: Agamidae) in south-eastern Australia'. *Australian Journal of Zoology* 47: 551–64.

Stuart-Fox, D. M., Hugall, A. F. and Moritz, C., 2002. 'A molecular phylogeny of rainbow skinks (Scincidae: *Carlia*): Taxonomic and biogeographic implications'. *Australian Journal of Zoology,* 50: 39–51.

Swan, G., 2017. *Reed Concise Guide: Snakes of Australia.* Reed New Holland, Sydney.

Swan, M. 2020. *Frogs and Reptiles of the Murray Darling Basin. A Guide to their Identification, Ecology and Conservation.* CSIRO Publishing. Clayton South, Vic.

Torr, G., 2000. *Pythons of Australia.* Australian Natural History Series. UNSW Press, Sydney.

Vanderduys, E., 2016. A new species of gecko (Squamata: Diplodactylidae: *Strophurus*) from north Queensland, Australia. *Zootaxa* 4117 (3): 341-358.

Vanderduys, E., 2017. A new species of gecko (Squamata: Diplodactylidae: *Strophurus*) from central Queensland, Australia. *Zootaxa* 4347 (2): 316-330.

Vanderduys, E., Hoskin, C. J., Kutt, A. S., Wright, J. M. and Zozaya, S., 2020. Beauty in the eye of the beholder: a new species of gecko (Diplodactylidae: *Lucasium*) from inland north Queensland, Australia. *Zootaxa* 4877 (2): 291–310.

Weigel, J., 1990. *Australian Reptile Park's Guide to Snakes of South-east Australia.* Australian Reptile Park, Gosford, NSW.

Williams, C. and Maier, C. (Eds) 2019. *A Tribute to the Reptiles and Amphibians of Australia and New Zealand.* Australian Herpetological Society. Reed New Holland, Sydney.

Wilson, S., 2003. *Reptiles of the Southern Brigalow Belt.* World Wildlife Fund, Australia.

Wilson, S. K., 1997. 'New information on *Pseudechis papuanus* (the Papuan Black Snake), a medically significant addition to Australia's reptiles'. *Memoirs of the Queensland Museum* 42(1): 232.

Wilson, S. K. 2012. '*Australian Lizards – A Natural History.*' CSIRO Publishing, Collingwood, Vic.

Wilson, SK., 2016. *Reptiles of the Scenic Rim.* Scenic Rim Regional Council.

Wilson, S. K. 2020. *Reed Concise Guide: Lizards of Australia.* Reed New Holland, Sydney

Wilson, S.K. and Knowles, D.G., 1988. *Australia's Reptiles. A Photographic Reference to the Terrestrial Reptiles of Australia.* Collins, Sydney.

Wilson, S. K and Swan, G. 2021. *A Complete Guide to Reptiles of Australia.* Sixth Edition. Reed New Holland, Sydney

Zug, G. R., Vitt, L. J. and Caldwell, J. P., 2001. *Herpetology. An Introductory Biology of Amphibians and Reptiles* (2nd edn). Academic Press, CA, USA.

INDEX TO SCIENTIFIC NAMES

INDEX TO COMMON NAMES